Praise for
The Mad, Mad, Mad World of Climatism

"Goreham, the antidote for Gore!"

— Doug Giles, Syndicated Radio Host

"This is the first book written to make you laugh at the absurdity of man-made global warming—that is, until it makes you cry. If 250 pages of facts are too much, you can simply read the amusing cartoons and quotes on every page to fully understand how the world has been misled."

— Jay Lehr, PhD, Science Director, The Heartland Institute

"It's all here. Replete with great graphics and much humor, Steve Goreham reveals the mountain of failed eco-predictions, rank green hypocrisies and outright fraud that currently masquerades in the guise of modern environmentalism and 'climate science'… *The Mad, Mad, Mad World of Climatism* is an easy-to-read, easy-to-understand expose of all the key facts behind the dirty 'green' politics of the greatest pseudo-science racket of our age."

— Peter Glover, International Associate Editor, *Energy Tribune*

"I am extremely impressed with this work, easily the best of its kind I have ever read… an authoritative, well-referenced, but easy-to-understand summary of the climate scare and its dire implications for society and the environment."

— Tom Harris, Executive Director, International Climate Science Coalition

"I strongly urge people to buy and read this book, and to keep it on hand for a quick reference."

— Russell Cook, Journalist, *American Thinker*

"Mad, bad, and dangerous to read!"

— Timothy Renshaw, *Business in Vancouver*

"Interesting, accurate, compellingly readable, and directly relevant to one of today's most important political debates. What more needs to be said beyond, 'Buy this fascinating book!'"

— Robert Carter, Australian Marine Geologist and Environmental Scientist

"This book is an impressive must-read work and is the best book on the issue of global climate change that I have ever read. It is an authoritative, well-referenced, and easy-to-understand work. It should be listed as required reading in all schools, universities, state legislative assemblies and, above all, the US Congress."

— James G. Speight, PhD, Editor, *Journal of Sustainable Energy Engineering*

"Steve Goreham's insightful and readable analysis of the ideology of Climatism and its socialistic goals is an important addition to public understanding of both the facts and the critical importance of making the right decisions for America."

— Harrison H. Schmitt, PhD, Geologist, former Senator and Apollo 17 Astronaut

"Steve Goreham has provided the science, the motivations, and the examples of deceit that surround the man-made warming hypothesis. He leaves no excuse for the public and the policy makers to prolong the misguided effort to spend trillions of dollars trying to reduce the insignificant effect of CO_2 on global climate change."

— Leighton Steward, Geologist, Author, and Environmentalist

"Once a belief such as human-caused climate change gains a critical mass of believers, it is difficult to stamp out—even when it's clearly wrong. This book aims to change the belief of everyday people."

— T. Dan Tolleson, PhD, *New American Standard*

"An amusing and colorful, yet science-based, look at mankind's obsession with global warming."

— *Publishers Weekly*

"Steve's new book is exactly what I hoped would be published. It makes climate science understandable to everyone. Finally, some common sense, presented in a way to inform, not to deceive."

— Burt Rutan, Aerospace Engineer, Innovator, Entrepreneur

"Goreham's book is an excellent, readable, comprehensive, and indispensable education for everyone. It should be required reading in all schools, universities, statehouses and Congress."

— Edwin Berry, PhD, Physicist and Meteorologist

"*The Mad, Mad, Mad World of Climatism* will be well-received by those skeptical of both climate doomsayers and the Obama administration."

— Alan Wallace, *Pittsburgh Tribune-Review*

"This book is useful and entertaining, and I recommend it."

— Thomas B. Fowler, ScD, Professor of Engineering, George Mason University

"Steve Goreham has made a major contribution to countering the unjustified alarmism about so-called 'anthropogenic global warming' and for destroying the myth about the 'science being settled…'"

— Terry Dunleavy, MBE, JP, New Zealand Entrepreneur

"…a very entertaining and informative book, *The Mad, Mad, Mad World of Climatism*, that I would heartily recommend that everyone read…"

— Alan Caruba, *Bookviews*

The
Mad, Mad, Mad World
of Climatism

Mankind and Climate Change Mania

STEVE GOREHAM

Steve Goreham (signature)

New Lenox Books

The
Mad, Mad, Mad World
of Climatism

Mankind and Climate Change Mania

© 2012 by Steve Goreham
www.climatism.net

ISBN: 978-0-9824996-2-7
Library of Congress Control Number: 2012907761

New Lenox Books, Inc.
New Lenox, IL USA 60451
newlenoxbooks@comcast.net

Printed in China

Credits:
Polar bears in convertible image by Canaan Shaffner
Burning wind turbine image by Polizei Stade
Cover Design: Peri Poloni-Gabriel, Knockout Design, www.knockoutbooks.com
Image of Harrison Schmitt on the moon from NASA
Cartoons by Bob Lynch

CONTENTS

FOREWORD

The dogmatic beliefs at the core of the ideology of "Climatism" lead its adherents to advocate increased government restrictions on human liberty, even on use of the most economical and environmentally benign sources of energy. Steve Goreham's insightful and readable analysis of this new ideology and its socialistic goals is an important addition to public understanding of both the facts and the critical importance of making the right decisions for America. If the climate alarmists weren't so serious about taking away our liberty in the name of controlling the climate, they would be funny.

Consider the following points:

1. Climate alarmists admit that for 4.5 billion years until about 50 years ago, natural events controlled Earth's climate. But then they claim that, all of a sudden, human produced carbon dioxide (CO_2) took over that control.

2. Climate alarmists predict that global temperatures will rise due to human use of fossil fuels, but they ignore the fact that temperatures have been rising naturally about one degree every century since the Little Ice Age began to wane 350 years ago.

3. Climate alarmists admit that strong correlations exist between variations in solar activity and variations in climate, but since they do not know a mechanism that would relate one to the other, they say that those correlations are meaningless.

4. Climate alarmists see a strong correlation in ice cores between increases in global temperature and increases in atmospheric CO_2, but they find it unimportant that the increases in CO_2 follow increases in temperature rather than precede them.

5. Climate alarmists consider climate to be controlled by CO_2 in the atmosphere, but they conveniently forget the oceans' capacity to absorb and release heat and CO_2 dwarfs that of the atmosphere by factors of 1,100 and 50, respectively.

6. Climate alarmists see current increases in atmospheric CO_2 as caused by the burning of fossil fuels, but they discount the fact that isotopic analysis shows that only four percent of the CO_2 in the atmosphere has come from fossil fuels with the vast majority being released from other sources, particularly from the oceans.

7. Climate alarmists believe that global warming and increases in atmospheric CO_2 will sound the death knell for human civilization but find it convenient to forget that warming is better for human longevity, and that increased CO_2 will significantly increase the food supply.

8. Climate alarmists believe that global climate models can predict the Earth's climate future but dismiss the fact that their predictions for atmospheric warming over the last 50 years have been proven wrong by direct measurement from satellites.

Nothing in the Constitution of the United States gives the Congress or the Executive Branch the power to attempt the task of regulating climate, as impossible as that would be under any realistic scenarios. No national security emergency exists relative to climate that would warrant increased governmental control of energy production. Today's Americans have an obligation to future Americans to elect leaders who do not believe in an omnipotent government but believe, as did the Founders, in limited government, and in the preservation of liberty and the natural rights of the people.

—Harrison H. Schmitt, PhD,
former US Senator and
Apollo 17 Astronaut

ACKNOWLEDGEMENTS

For twenty years, since the 1992 Earth Summit, realist climate scientists have challenged the rising ideological philosophy that man-made emissions are causing catastrophic global warming. Ridiculed, scorned, labeled "deniers," and called stooges for big oil and coal, these courageous men and women have taken the blows and stood firm for proper climate science and against ideological alarmism. This book references the scientific work and opinions of Dennis Avery, Bob Carter, Joseph D'Aleo, David Evans, Bill Gray, William Happer, Craig Idso, Dick Lindzen, Stephen McIntyre, Ross McKitrick, Joanne Nova, Fred Singer, Roy Spencer, Henrik Svensmark, Anthony Watts, and many other realist climate scientists. *The Mad, Mad, Mad World of Climatism* stands on the shoulders of these heroes. Special mention goes to Anthony Watts and his blog *WattsUpWithThat*, the world's most informative climate science website. This site provided me with an endless stream of articles about our misguided mankind, caught in the grip of climate madness. Craig Idso's excellent website, *CO$_2$Science*, provided me with studies and graphs on historical temperature trends.

Special thanks goes to Joe Bast, Ed Berry, Bob Carter, Don Dears, John Droz, Tom Harris, Howard Hayden, Norm Rogers, and Leighton Steward for review and correction of the text. I'm in the debt of Harrison Schmitt for adding his comments in the Foreword. Thanks also to my book consultant Peri Poloni-Gabriel and my editors Helen Chang and Janet Weber, for their professionalism and patience with me. The image of cruising polar bears was developed by Canaan Shaffner and the great cartoons came from Bob Lynch.

Oh, and let's not forget Al Gore, Tim Flannery, James Hansen, Michael Mann, Bill McKibben, Nicholas Stern, David Suzuki, and many others who are overly concerned about global warming. Without their sensational warnings of alarm about the coming climate catastrophe, this book would not be necessary.

This book is dedicated to my wife, Sue, the love of my life, who has endured my many months of toil and endless conversations about global warming, and also my family and friends for their valuable inputs and inspiration.

INTRODUCTION

'll bet I know your thoughts about the climate. For years you've heard about how Earth is warming up. How people are the cause of global warming. How the polar bears are threatened with extinction. How we each must change our lifestyle for the good of the planet.

Television specials show calving glaciers and raging torrents from an ice melt in Greenland and voice concern over greenhouse gas emissions. Scientists report from Antarctica about pending disasters. A news story says that the flood in Pakistan is due to global warming. And wasn't Hurricane Katrina caused by climate change?

If you listen to the news, your national leaders promote new policies to fight climate change. Your nation must embrace renewable energy and reduce greenhouse gas emissions. There is talk about new taxes and regulations that will require sacrifices, but these are necessary to solve the climate crisis.

Of course, as a good citizen, you try to follow the lead. You've purchased some of the new compact fluorescent lights. They're a little expensive and it takes a while for them to get bright. They contain mercury—so you don't want to break one. Is it true that you can't buy any of the old incandescent bulbs anymore?

You're told that electric cars are the hot new technology. But they seem a little small and are said to have only a 40-mile range. Will they be available as a minivan or a pickup truck? If you buy one, where can you charge it?

You might have a new Vice President of Sustainability at your company. Purchases of expensive green energy and estimating the carbon dioxide output from processes are new policies. It's politically incorrect to question these policies, so you remain silent.

Your high school student comes home with concerns about climate change. It seems

1

she has just seen Al Gore's movie in class. She asks if your family is doing enough to help save the planet.

A group of wind turbines was recently constructed in the next county. They look majestic, towering above fields and grazing livestock. But when you drive past them, many seem to be standing idle.

Yes, the world is certainly a greener place in response to all these changes. Yet, something deep down in your gut says that all this alarm about global warming just doesn't ring true. Maybe you've heard the demands for change, but they don't make sense in your daily life. Maybe you remember the 1970s, when scientists were concerned about global cooling and a pending ice age. But friends tell you now that your memory is faulty—there was no fear of an ice age back then.

You've been told that our air is being filled with "dangerous carbon pollution." But, you don't see any evidence of this. You recall the smog in our cities and foul-smelling polluted air when you were a child. Somehow it seems like the quality of air has improved during the last 30 years, despite the alarms from the news media.

Maybe you've just been through a tough winter, with mountainous drifts of snow and cold temperatures. Didn't the seasonal forecast call for a warm, dry winter? And what about Climategate—something about a scandal over temperature data at a university in Britain?

Well, your intuition about global warming is right. There is no direct scientific evidence that man-made greenhouse gases are causing catastrophic global warming. Instead, the world has been captured by the ideology of Climatism— the belief that man-made greenhouse gases are destroying Earth's climate. Most of the leaders in government, at universities, in scientific

FAILED PREDICTIONS

"There are ominous signs that the earth's weather patterns have begun to change dramatically and that these changes may portend a dramatic decline in food production...after three quarters of a century of extraordinarily mild conditions, the earth's climate appears to be cooling down."
—"The Cooling World"
Newsweek, April 28, 1975[1]

BEWARE DANGEROUS CARBON POLLUTION!

organizations, and in business say they believe in Climatism.

The astonishing thing is that *CO_2 is green*! Rather than being a pollutant, carbon dioxide makes plants grow! In a world turned upside down, every community and every company measures their "carbon footprint" and tries to reduce emissions of a harmless, invisible gas that is essential for photosynthesis and the growth of plants.

Don't misunderstand me. There are real pollutants that we need to control. For more than 25 years, I've had the joy of kayaking many of the great white water rivers of North America. From Texas to Idaho to Quebec, rivers have been a love of my life. I've paddled creeks on the Cumberland Plateau in Tennessee, Al Gore's home turf. Rivers are highlights of this amazing and beautiful world. We all want our water to be pure and our air to be clean. We're all environmentalists. But we must use sound science to determine man-made impacts on our climate. Sensible economics should drive our energy policy, not unfounded fears about global warming.

This book will take a common-sense look at global warming mania. We'll provide a down-to-earth discussion of the science, which increasingly shows that natural cycles of Earth are the dominant cause of climate change—not man-made greenhouse gas emissions. We'll discuss how climate science has been corrupted and look at the money and special interests that continue to drive the dogma of Climatism forward. We'll discuss renewable energy, which is proposed as a primary solution to stop climate change.

The arguments of this book are not just opinions, but are based on the work of

The author at play on the Peshtigo River in Wisconsin.
Photo by Guenther[2]

Big Green for Climate Change

EPA Awards $17 Million to Support Research on the Impacts of Climate Change
"Twenty-five universities to explore public health and environmental facets of climate change"
—Environmental Protection Agency news release, Feb. 17, 2010[3]

hundreds of scientists across the world who challenge the theory of man-made global warming. Graphs and scientific data from peer-reviewed papers are used to show that man-made influences are actually only a very small part of Earth's climate. The evidence is available for all to see.

Chapters 1–2 discuss how our leaders have been captured by the false ideology of Climatism and the remedies proposed to change the life of every person on Earth. Climate science is discussed in Chapters 3–5. I encourage you to read the down-to-earth science in these chapters, but of course feel free to skip these if you're just interested in how the world has been smitten by climate madness. Chapters 6–7 discuss alarming claims about Earth's icecaps and weather, and show that these claims are not supported by scientific data. Chapter 8 exposes some of the biggest whoppers of global warming mania. Chapters 9–10 discuss bad science and the powerful role that money plays in this whole affair. Chapter 11 discusses the continuing shortcomings of renewable energy, despite many decades of media hype and promotion by governments. Don't miss Chapter 12, "You Can't Make this Stuff Up!"

Climate change is a serious topic. Government policies are proposed or already in place that will affect the light bulbs you buy, the construction of your house, the car or appliance you purchase, the price of your energy, your workplace, what your children are taught in school, and almost every aspect of your life. This book will help you sort fact from fiction in the global warming debate. It will remove the fear and paranoia that you and your family may be feeling from daily bombardment of climate change nonsense from work, school, and community.

Along the way, we'll have some fun. We'll discuss the wackiness of mankind turned on its head by global warming alarmism. This whole charade has moved from the serious to the absurd. Beware the sidebars, since a few of these are spoofs. But the rest are true headlines or quotes from our mad, mad, mad world of Climatism. Enjoy and, as the great Paul Harvey used to say, learn "the rest of the story" about climate change.

Steve Goreham

CHAPTER 1

MANKIND IN THE GRIP OF A MADNESS

"According to a new UN report, the global warming outlook is much worse than originally predicted. Which is pretty bad, when they originally predicted it would destroy the planet."—COMEDIAN JAY LENO

The scene was the Grand Hotel in Oslo, Norway, December 10, 2007. The Norwegian Nobel Committee awarded the Nobel Peace Prize to Albert Arnold Gore, Jr. along with the Intergovernmental Panel on Climate Change (IPCC):

…for their efforts to build up and disseminate greater knowledge about man-made climate change, and to lay the foundations for the measures that are needed to counteract such change…[1]

Former Vice President Gore, the leading doomsayer for man-made global warming, delivered a rousing, alarming acceptance speech:

5

Al Gore at Nobel Prize acceptance speech.
(Photo by Bjornsrud, 2007)[2]

We, the human species, are confronting a planetary emergency—a threat to the survival of our civilization that is gathering ominous and destructive potential…the earth has a fever. And the fever is rising…Indeed, without realizing it, we have begun to wage war on earth itself…[3]

The IPCC is an organization of the United Nations that was formed in 1988. According to the IPCC, its purpose is:

…to provide the world with a clear scientific view on the current state of knowledge in climate change and its potential environmental and socio-economic impacts.[4]

From 1990 to 2007, the IPCC compiled four assessment reports with hundreds of scientists participating. Recent reports were huge documents of over a thousand pages. Each successive report was more certain that humans are responsible for global warming.

Rajendra Pachauri, Chairman of the IPCC, accepted the award on behalf of the organization. During his speech, Pachauri pointed to the technical material in the assessment reports, which form the scientific basis for the theory of man-made climate change that is accepted by most of the world. He quoted the Fourth Assessment Report of 2007:

…warming of the climate system is unequivocal…most of the global average warming over the past 50 years is very likely due to anthropogenic [man-made] greenhouse gases increases…[5]

The 2007 Nobel Prize crowned 20 years of building the delusion of man-made global warming. Al Gore and the IPCC shared the Nobel monetary award of 10 million Swedish Kroner ($1.6 million), and the world was turned upside down over Climatism.[6]

Climatism is the belief that man-made greenhouse gases are destroying Earth's climate. As we'll see throughout this chapter, most governmental, scientific, and business leaders of the world say they accept the ideology of Climatism. But as we'll show in Chapters 4 and 5, an increasing body of science shows that climate change is due to natural processes, probably driven by the sun, and that man-made emissions play only a very small part.

MEET THE CLIMATE PROPHETS OF DOOM

Al Gore is the best-known spokesman, but only one of many claiming that Earth's climate is going to Hades. Gore is author of three books on climate, including the best-selling *An Inconvenient Truth*, which was made into an Academy Award-winning documentary in 2006.[7] The cover of his latest book, *Our Choice*, shows an image of North America from space with a massive tropical storm approaching California from the Pacific Ocean.[8] The tropical storm was computer generated because a real photo could not be found with a big enough storm.

In 2009, Gore warned the US Senate:

> We have arrived at a moment of decision. Our home—Earth—is in danger. What is at risk of being destroyed is not the planet itself, of course, but the conditions that have made it hospitable for human beings.[9]

Dr. James Hansen is regarded as a leading scientific voice for the theory of man-made warming. Dr. Hansen is director of the NASA Goddard Institute for Space Studies in New York and Adjunct Professor of Earth Sciences at Columbia University. Dr. Hansen proposes that should Earth's surface temperature rise by only one degree, the icecaps will melt and the seas will rise 20 feet. He argues that Earth's climate will pass a "tipping point," causing extreme changes regardless of any measures mankind could take:

> The climate is nearing tipping points. Changes are beginning to appear and there is a potential for explosive changes, effects that would be irreversible, if we do not rapidly slow fossil-

California believes in tipping points!
(Image spoof by Watts)[10]

THE LEADING CLIMATE ALARMISTS[11]

"...unless we act boldly and quickly to deal with the underlying causes of global warming, our world will undergo a string of terrible catastrophes..."
—*Al Gore (2007)*[12]

"Burning all the fossil fuels will destroy the planet we know, Creation, the planet of stable climate in which civilization developed."
—*James Hansen (2008)*[13]

"The world hasn't ended, but the world as we know it has—even if we don't quite know it yet."
—*Bill McKibben (2010)*[14]

"Climate change is a result of the greatest market failure the world has seen...We risk damages on a scale larger than the two world wars of the last century."
—*Nicholas Stern (2007)*[15]

"One problem facing humanity is now so urgent that, unless it is resolved in the next two decades, it will destroy our global civilization: the climate crisis."
—*Tim Flannery (2009)*[16]

fuel emissions over the next few decades. As arctic sea ice melts, the darker ocean absorbs more sunlight and speeds melting. As the tundra melts, methane, a strong greenhouse gas, is released, causing more warming. As species are exterminated by shifting climate zones, ecosystems can collapse, destroying more species…The greatest danger hanging over our children and grandchildren is initiation of changes that will be irreversible on any time scale that humans can imagine.[17]

Dr. Hansen has acted as the lead scientific advisor for much of the ideology of Climatism. His 1981 paper on global warming provided the basis for the IPCC concept of "radiative forcings," and the development of a range of temperature-increase forecasts based on different levels of energy usage.[18] Dr. Hansen has been a close advisor to Al Gore and supplied technical advice for *An Inconvenient Truth*. Most recently, Dr. Hansen recommended to Bill McKibben to choose *350.org* as the name for his activist global warming group.

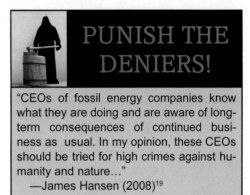

PUNISH THE DENIERS!

"CEOs of fossil energy companies know what they are doing and are aware of long-term consequences of continued business as usual. In my opinion, these CEOs should be tried for high crimes against humanity and nature…"
—James Hansen (2008)[19]

Environmentalist Bill McKibben is a star on the list of climate doomsayers. McKibben is a journalist and scholar at Middlebury College in Vermont, and author of several books on the climate. His most recent, *Eaarth: Making a Life on a Tough New Planet*, claims that our world has already been irreversibly changed by human emissions.[20] In 2008 McKibben founded *350.org*, a group focused on grassroots efforts to convince people of man-made climate change. The name of the organization is taken from 350 parts per million, the maximum safe level of atmospheric CO_2 according to James Hansen. In October 2009, *350.org* coordinated 5,200 rallies in over 181 countries[21] to send a message to "solve the climate crisis." According to McKibben:

> There's no happy ending where we prevent climate change any more. Now the question is, is it going to be a miserable century or an impossible one, and what comes after that.[22]

Lord Nicholas Stern is one of the leading climate sirens of the United Kingdom. Lord Stern is Professor of Economics and Government at the London School of Economics. In 2006, he led a team that published the *Stern Review: The Economics of Climate Change* for then Prime Minister Tony Blair. The report projected economic losses of 20 percent of the global economy by the year 2100 for a "business-as-usual" scenario. The report recommended that the world annually spend one percent of global Gross National Product (GNP) to reduce emissions.[23] In 2008, Stern upped the needed cost to two percent of world GNP because he said climate change was happening faster than previously thought.[25]

"...what we are talking about then is extended world war...People would move on a massive scale. Hundreds of millions, probably billions of people would have to move..." —Nicholas Stern (2009)[24]

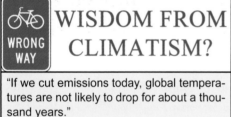

WISDOM FROM CLIMATISM?

"If we cut emissions today, global temperatures are not likely to drop for about a thousand years."
—Tim Flannery (2011)[26]

Dr. Tim Flannery, professor at Macquarie University in Sydney, is the leading climate siren from the Southern Hemisphere. He's been called the "Al Gore of Australia." Dr. Flannery wrote the best-selling book *The Weather Makers*, which has sold more than a million copies. According to Flannery, we can now control Earth's temperature:

> When we consider the fate of the planet as a whole, we must be under no illusions as to what is at stake. Earth's average temperature is around 59°F, and whether we allow it to rise by a single degree or 5°F will decide the fate of hundreds of thousands of species, and most probably billions of people.[27]

Warning about the coming climate catastrophe has been an effective path to fame and fortune for these leaders and many others. For example, Mr. Gore receives more than $100,000 for every speech to audiences about climate change. His net worth has increased from a few million at the end of his term as Vice President to more than $100 million today.[28] The simple message of "greenhouse gas emissions produce global warming, produces climate catastrophe" is plausible, and has been readily accepted by the public. But as we'll show in Chapters 4 and 5, a closer look at the science shows that these climate pied pipers are leading us astray.

WORLD LEADERS CAPTURED BY CLIMATISM

The leaders of the world have been captured by the ideology of Climatism. Today, 193 of 194 heads of state say they accept the theory of man-made global warming and are pursuing policies to stop our planet from warming. The sole exception is Vaclàv Klaus, President

THE EMPEROR HAS NO CLOTHES[29]

"We're all agreed that climate change is one of the greatest and most daunting challenges of our age. We have a moral imperative to act and act now."
—*UK Prime Minister David Cameron (2010)[30]*

"Few challenges facing America and the world are more urgent than combating climate change. The science is beyond dispute and the facts are clear."
—*US President Elect Barack Obama (2008)[31]*

"...climate change is real and it is caused to a significant extent by human activity."
—*Australia Prime Minister Julia Gillard (2010)[32]*

"...climate change is accelerating. It threatens our well being, our security, and our economic development. It will lead to uncontrollable risks and dramatic damage if we do not take resolute countermeasures..."
—*German Chancellor Angela Merkel (2009)[33]*

"India was a late comer to industrialization, and as such, we have contributed very little to the accumulation of greenhouse gas emissions that cause global warming. But we are determined to be part of the solution to the problem."
—*India Prime Minister Mahoman Singh (2009)[34]*

Signs that
Man-Made Warming
is Happening???

"Global warming causes volatility. I feel it when I'm flying."
—US Senator Debbie Stabenow (2009)[35]

of the Czech Republic, who is an outspoken critic of climate alarmism.

Almost all of the major universities of the world, the leading scientific organizations, and most Fortune 500 companies agree that humans are to blame for global warming. The media and Hollywood beat the drums for lifestyle changes to save the planet. Pew Charitable Trusts estimates that in 2010 the world spent $243 billion to adopt renewable energy in an effort to reduce carbon dioxide emissions, a 30 percent increase over 2009.[36]

And it's not just at the national level. Your mayor or burgomaster is likely planning green restrictions on your house, your company, and your school. Your official will say that if everyone would just install solar cells on their roof and buy electricity from windmills, we can stop global warming. A Climate Protection Agreement has been signed by more than 1,000 US mayors, representing a population of over 100 million, pledging to enact policies to "meet or beat the greenhouse gas emissions target suggested for the United States in the Kyoto Protocol."[37]

The C40 Cities Climate Leadership Group is an association of the mayors of large cities "committed to tackling climate change." C40 membership includes 40 of the world's largest cities such as Athens, Buenos Aires, and Tokyo, and a growing number of affiliated cities. The C40 holds summit meetings every two years for mayors, their staff, and business leaders to meet to "share best practices and identify collaborative projects all aimed at tackling climate change."[38]

American students are being taught climate propaganda at all school levels. In elementary and middle schools they hear that if we change our light bulbs, we can save polar bears. Environmental science is now a popular course in high schools and universities, with climate change a key topic. As of July, 2011, the presidents of 676 US universities and colleges had signed the "President's Climate Commitment." The commitment states:

> We recognize the scientific consensus that global warming is real and is largely being caused by humans. We further recognize the need to reduce the global emission of greenhouse gases by 80% by mid-century at the latest, in order to avert the worst impacts of global warming and to reestablish the more stable climatic conditions that have made human progress over the last 10,000 years possible.[39]

CLIMATE CHANGE IS NOW BIG BUSINESS

Kevin Parker, Director of Global Asset Management at Deutsche Bank, talked about Al Gore and the company's green investment fund:

> He [Al Gore] impressed us all at Deutsche Bank Asset Management. We invited him to an internal meeting in April 2007 during which we discussed the issue of climate change extensively. A few months later, he received the Nobel Peace Prize for his commitment. We then created a fund that invests in companies that position themselves as climate-neutral. **Within two months almost 10 billion dollars flowed into this fund. Can you imagine? 10 billion! There has never been such an overwhelming success.** [Emphasis added][40]

There's big money in saving the planet. Companies now have a financial motive to promote green business, even if the theory of man-made global warming is wrong. Deutsche Bank is just one of thousands of firms that profit from selling green products, supplying renewable energy, trading carbon credits, or providing consulting or legal advice on climate change. In addition to over $200 billion that nations spend annually on renewable energy, the global green market for consumer goods is now hundreds of billions per year and growing strongly. According to on-line retailer Kelkoo, green product sales in Europe reached €56 billion ($74 billion) in 2009.[41] According to the Natural Marketing Institute, US consumers spent about $300 billion in 2008 for LOHAS (Lifestyles of Health and Sustainability) products.[42] Much of this spending was for natural and organic products, but a growing percentage was for products with a "small carbon footprint," including consumer goods, housing, eco-tourism, and alternative transportation. As good eco-citizens, consumers are willing to pay a considerable price premium for green products over alternatives.

WACKY CLIMATE CHANGE HEADLINES

Russian Scholar Warns of 'Secret' US Climate Change Weapon
—*Radio Free Europe Radio Liberty,* July 30, 2010[43]

The investment dollars involved in climate alarmism are huge. In October 2011, prior to the opening of the Durban Climate Conference, a coalition of 285 investor groups holding over $20 trillion in assets issued a joint statement calling for:

> …policy action which stimulates private sector investment into climate change solutions, creates jobs, and is essential for ensuring the long-term sustainability and stability of the world economic system.[44]

In other words, agree to a global climate treaty so we can make money in renewable energy and biofuels. Among the signers were investment heavyweights such as BT Financial Group, Deutsche Bank, HSBC, and Sumitomo Bank.

The US Climate Action Partnership is an association of more that 20 companies and organizations that has been lobbying for cap-and-trade and other climate legislation in the United States. Membership has included a Who's Who of multinational corporations, including DuPont, Ford, General Electric, General Motors, Honeywell, Shell, and Siemens.[45] The combination of governmental policy, pressure from environmental groups, the desire to be good corporate citizens, and the opportunity to earn green profits has captured almost all major global corporations. Even the American Petroleum Institute has officially adopted the theory of man-made global warming.

SCIENCE SAYS

Climatism relies heavily on the mantle of science authority. The IPCC claims that 2,500 scientists wrote their Fourth Assessment Report. A survey published in 2009 claimed that 97 percent of "climatologists who are active publishers on climate change" believe that human activity is significantly changing global temperatures.[46] Despite many flaws, the survey is still widely quoted in the news media.

In fact, all major scientific organizations of the world have taken the official position that humankind is causing global warming. But, although the governing boards of these groups endorse man-made climate change, the members of each organization do not necessarily agree. Indeed, climate science has been corrupted. As we'll discuss later, financial pressures have led major scientific organizations to adopt Climatism.

HOW DID WE GET TO THIS STATE OF AFFAIRS?

We're in the midst of the greatest global delusion in history. As we'll soon discuss, mounting scientific evidence shows that natural factors are the cause of global warming, not man-made greenhouse gas emissions. How did we get to our current situation, where the world's governments, universities, businesses, and scientific organizations profess man-made global warming?

In fact, the world jumped to a conclusion. In June 1988, Senator Tim Wirth, then chair of the Committee on Energy and Natural Resources, held the first-ever hearing on the science of climate change. The star testimony came from Dr. James Hansen, who stated that he was:

Major Scientific Organizations Supporting Man-Made Warming

American Physical Society (APS):
"The evidence is incontrovertible: Global warming is occurring."[47]

National Oceanic and Atmospheric Administration (NOAA):
"...human activity is the primary driver of recent warming."[48]

National Aeronautic and Space Administration (NASA):
"The current warming trend...is very likely human induced..."[49]

American Geophysical Union (AGU):
"...emissions of CO_2 must be reduced by more than 50 percent..."[50]

American Meteorological Society (AMS):
"...human activities are a major contributor to climate change..."[51]

National Academy of Sciences (NAS):
"...need for substantial action to limit the magnitude of climate change..."[52]

Royal Society (UK):
"...evidence that the warming...has been caused largely by human activity..."[53]

Royal Meteorological Society (RMetS, UK):
"Climate change is now one of the major hurdles facing the global community..."[54]

European Academy of Sciences and Arts (EASA):
"Human activity is most likely responsible for climate warming."[55]

Canadian Meteorological and Oceanographic Society (CMOS):
"Human activities must...significantly reduce emissions starting immediately."[56]

"...99 percent confident that the world really was getting warmer and that there was a high degree of probability that it was due to human-made greenhouse gases."[57]

Dr. Hansen's testimony, an unusually hot summer, and the convening of the UN Conference on the Changing Atmosphere in Toronto that same month boosted media attention on global warming.

Later in 1988, the United Nations established the

"I'm 99 percent confident..."
—James Hansen, 1988
Senate testimony[58]

IPCC. The UN intended to lead the world in environmental policy and the IPCC would play a large role in this effort. In 1990, the IPCC issued its First Assessment Report, finding that human emissions were significantly raising global temperatures. Then at the 1992 Earth Summit, the United Nations Conference on Environment and Development in Rio de Janeiro, 41 nations and the European Community (EC) signed the Framework Convention on Climate Change, a treaty vowing to:

> …adopt national policies and take corresponding measures on the mitigation of climate change, by limiting…emissions of greenhouse gases.[59]

Many scientists disagreed with the findings of the IPCC and the conclusion of the Rio Earth Summit and expressed disagreement by signing a petition.[60] But the IPCC and the United Nations, the EC, advocating nations such as the United Kingdom and Germany, the news media, environmental groups, and scientists using climate computer models won the field. Climatism gained momentum and was reinforced by the signing of the Kyoto Protocol in 1997, when 38 nations agreed to mandatory reductions in greenhouse gas emissions.[61]

For the last 20 years since the Rio Earth Summit, the world's political leaders have been arguing about how much to reduce emissions and how quickly. Yet, more and more science now shows that natural cycles of Earth, driven by the sun, are the primary cause of recent global warming. But the climate movement is like a laden cargo ship, propelled forward with huge momentum. It will take many years to turn it around.

MANKIND IN THE GRIP OF A MADNESS

If madness can be defined as being out of touch with reality, that's where we are. The idea that a small increase in a trace gas in our atmosphere can cause snowstorms in New York, heat waves in Europe, and polar bear extinction doesn't hold up after a closer look at the science. The idea that we can stop the rise of the seas by changing our light bulbs is ludicrous.

But as we'll discuss in the next chapter, our world leaders are implementing all sorts of policies to try to solve a problem that doesn't exist. It's full speed ahead to fight global warming. Buckle your seat belt for this stranger-than-fiction story.

CHAPTER 2

DO THIS TO SAVE THE PLANET

"The planet has been through a lot worse than us…been through earthquakes, volcanoes, plate tectonics, continental drift, solar flares, sun spots, magnetic storms, the magnetic reversal of the poles, hundreds of thousands of years of bombardment by comets and asteroids and meteors, worldwide floods, tidal waves, worldwide fires, erosion, cosmic rays, recurring ice ages—and we think some plastic bags and some aluminum cans are going to make a difference?"
—COMEDIAN GEORGE CARLIN

So, what's so bad about working together to try to keep the planet from warming up? Can't everyone switch to fluorescent lights and use a little less energy? These are reasonable questions, but they hardly scratch the surface of what Climatists have in store for us. Nancy Pelosi, former Speaker of the US House of Representatives, discussed personal impacts of climate change:

We have so much room for improvement. Every aspect of our lives must be subjected to an inventory…of how we are taking responsibility.[1]

Not only your light bulbs, but your vehicle, your house, your business, your diet, your family, and your government must all change if we are to have any hope of stopping global warming.

BACK TO THE HORSE & BUGGY

More than 15,000 persons from over 190 nations met in Cancun, Mexico, during December of 2010 to once again discuss the problem of climate change and what to do about it. A UN climate conference is now an annual event in which thousands journey to scenic locations to discuss the twenty-first century's hottest topic. While the Cancun meeting produced no legally binding agreements on measures to reduce emissions, the delegates rededicated their efforts to reduce greenhouse gases to "hold the increase in global average temperature below 2°C above pre-industrial levels."[2] To achieve this, world leaders are calling for severe reductions in emissions by the developed countries.

The European Union has committed to an 80 percent reduction in emissions by mid-century. President Barack Obama agrees, pledging to reduce US emissions by 83 percent by 2050 from 2005 levels.[3] This is quite a promise. US emissions in 2005 were 5.4 tons of carbon per person each year. This is actually about carbon dioxide, which is produced by almost all of our industrial processes, but "carbon" is used as the measure.

Back in 1960, when most families had a television and commercial jet aircraft became commonplace, per capita emissions were about 4.5 tons per year. By comparison, in 1920, when automobiles and washing machines were first purchased by many families, emissions were about 4 tons per year. In 1900, when electricity and light bulbs began to reach homes, emissions were about 2 tons of carbon per year. Since the US population is projected to increase to 439 million by 2050,[5] we need to go all the way back to 1870, down to a level of only 0.7 tons per person, to get an 83 percent reduction.

Nigeria emits about 0.7 tons of carbon per person per year.[6] However, half the population doesn't have access to electricity. Only 3 of every 100 Nigerians drive a car. Very green. The

WISDOM FROM CLIMATISM?

"A massive campaign must be launched to de-develop the United States. De-development means bringing our economic system (especially patterns of consumption) into line with the realities of ecology and the world resource situation."
—Paul Ehrlich, Anne Ehrlich, and John Holdren (1970)[4]

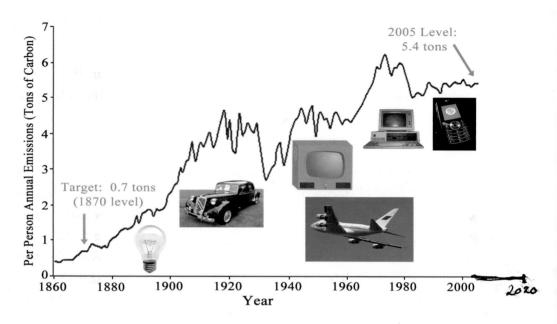

US per Capita Emissions, 1860–2005. The US government is calling for an 83 percent emissions reduction by 2050, equal to the level of per person emissions in 1870. (US Census Bureau, 2011; Carbon Dioxide Information Analysis Center, 2011)[7,8,9]

government's call for an 83-percent reduction is a plan to return to a horse-and-buggy nation. *Neither President Obama nor any other leader knows how to do this without destroying the world's economies.* Note that John Holdren, Director of the Office of Science and Technology Policy and President Obama's top science officer, called for the de-development of the United States in a book written in 1970.

RENEWABLE ENERGY: THE ILLUSORY SOLUTION

Renewable energy is proposed as the solution. The claim is that if we just stop using coal, natural gas, and oil and instead adopt solar, wind, and biofuels, we can stop "dangerous carbon emissions" and halt global warming. Unfortunately, there are as many misconceptions about renewable energy as there are about climate science. We'll discuss renewable energy in more detail in Chapter 11, but here are a few facts:

A. **Renewables are very expensive.** Onshore wind is more costly than traditional fuels and offshore wind is much more expensive for production of electricity. Solar is 2.5–4 times more expensive than traditional fuels.[10]

B. **Wind and solar are intermittent.** When the wind doesn't blow, hydrocarbon plants must instead supply power. Denmark and Germany have together built 25,000 wind turbines but have been unable to close a single coal-fired power plant.[11]

C. **Renewables provide only a pittance of energy.** After 20 years of subsidies, more than 160,000 wind turbine towers have been installed across the world.[12] Yet, in 2010 these turbines provided less than 1 percent of the world's energy. Hydrocarbons continue to provide over 80 percent of the world's energy.[13]

D. **Biofuels are a poor substitute for gasoline.** After accounting for all the energy required to grow corn and refine ethanol motor fuel, about 4.5 gallons of corn ethanol are needed to replace the energy equivalent of 1 gallon of gasoline.[14]

E. **Renewables don't significantly reduce greenhouse gas emissions.** Wind turbines must be backed up by hydrocarbon plants that cycle on and off to track the intermittency of the wind, eliminating nearly all emissions savings. Burning ethanol fuel from corn emits more carbon dioxide than gasoline.[15]

Because renewables can't compete cost effectively with hydrocarbon, nuclear, and hydropower sources of energy, governments must force adoption. Subsidies, feed-in tariffs, loan guarantees, and legislative mandates are all policies used to boost renewables. Wind and solar are subsidized 30 percent to 50 percent by world governments,[16] backed up by laws to force acceptance by utilities. Total direct and indirect subsidies for biodiesel and ethanol across the world are $1 to $2 per gallon (€0.20 to €0.40 per liter).[17]

But even these incentives aren't enough for climate alarmists. Efforts are underway to establish a "price on carbon" in the form of either a cap-and-trade carbon trading system or a direct carbon tax. By making coal, gas, and oil expensive, governments intend to artificially create a demand for renewables. Europe installed its Emissions Trading System (ETS) in 2005, the first major cap-and-trade system.

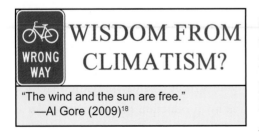

WISDOM FROM CLIMATISM?

"The wind and the sun are free."
—Al Gore (2009)[18]

The size of the "needed" transition to renewables is staggering. The world used more than 20,000 terawatt-hours of electricity[19] in 2010, and the 160,000 installed wind turbines provided only about 350 of those terawatt-hours,[20] or about 1.7 percent of the demand. About 3.5 billion of the world's people reside in developing nations. Their economic growth is expected to more than double world electricity demand during the next 40 years. For wind to provide just 15 percent of the world's electricity in 2050, about three million of the 400-foot-high towers would need to be built at more than $2 million apiece. Of course, conventional power plants must also be built to supply power when the wind doesn't blow.

This just isn't going to happen. There isn't enough money in the world to subsidize this charade, although wind turbine suppliers Vestas and General Electric would like us to try. Don't think that solar energy is a short-term answer, either. After 20 years of incentives, solar meets *less than one-tenth of one percent* of world or US needs. In the US, the highest demand for solar energy remains the heating of swimming pools.

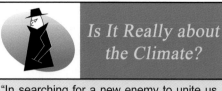

Is It Really about the Climate?

"In searching for a new enemy to unite us, we came up with the idea that pollution, threat of global warming, water shortages, famine and the like would fit the bill. All these dangers are caused by human intervention, and it is only through changed attitudes and behavior that they can be overcome. The real enemy then, is humanity itself."
—Alexander King, founder of the Club of Rome environmental think-tank, advisor to the United Nations (1991)[21]

SEVEN MAJOR LIFESTYLE CHANGES UNDERWAY

Renewable energy receives most of the money in the crusade to control climate change, but Climatism advocates have a host of other recommendations for each of us. These proposals won't impact global warming, since, as we'll show later, man-made emissions aren't the primary cause. But they will negatively impact the lifestyle of every citizen.

Your vehicle—electrify it!
According to the Sierra Club:

> Each gallon of gasoline we burn emits between 24 and 28 pounds of carbon dioxide—the most common greenhouse gas—into the atmosphere.[22]

As we'll discuss in Chapter 5, CO_2 is *not* nature's most common greenhouse gas. Nevertheless, climate alarmists argue that we need to drive fewer miles, use smaller cars, and reduce emissions from all vehicles. Consumers choose cars for functionality, style, roominess, safety, mileage, and driving performance. Businesses choose vehicles for business utility and performance versus operating cost. Until recently, no one purchased a car to reduce emissions. Yet, world governments are in the process of forcing us to change our vehicles to try to control climate change.

Since the oil price shock of the 1970s, developed nations have established policies to increase gas mileage. Today, cars of Japan and the European Union lead the world with an average fuel economy of over 40 miles per gallon (mpg), or 17 kilometers per liter, achieved through a combination of taxes, government mandates, and voluntary agreements with automakers. US and Canadian automobile fleets have the lowest fuel economy of developed countries, about 26 mpg and 33 mpg, respectively.[23] Reduction in oil imports to improve trade balances and to reduce oil dependency on rogue nations may be good reasons for boosting mileage. However, onerous regulations to reduce greenhouse gas emissions are now burdening world automobile industries.

In 2008, European automakers fell short of a goal that cars should emit no more than 140 grams of carbon equivalent per kilometer (g/km) driven. Later that year, the European Commission (EC) adopted a standard of 120 g/km, effective in the year 2015, with a goal of 95 g/km in 2020.[24] The 2008 target was voluntary, but in the future automakers will be fined for each vehicle produced if they are unable to meet the standard.

TNO Management Consultants, a consultant for the EC, estimates a retail price increase of €1,200 per car for the industry to reach 140 g/km and an additional increase of €2,450 to meet the new standard of 120 g/km. By 2015, European cars will be €3,650 ($4,850) more expensive due to emissions regulations.[25] Good citizens of Europe, enjoy your new car prices.

The US has been a laggard, but the Environmental Protection Agency (EPA), is moving aggressively to slap new regulations on vehicles. On April 1, 2010 the EPA proposed to boost the Corporate Average Fuel Economy (CAFE) standard from 27 mpg to 35.5 mpg and to introduce the first US emissions standard of 250 grams/mile (155 g/km). Both of these mandates would begin in 2016.[26] But this was just the beginning. On September 9 the same year, 20 environmental groups of the "Safe Climate Campaign" sent a letter to President Obama

CLIMATIST HIT LIST

The Sport Utility Vehicle

calling for CAFE standards of 60 mpg.[27] Efforts by the EPA and Department of Transportation, in discussion with automakers, culminated in November of 2011 in a fleet-wide CAFE standard of 54.5 mpg for automobiles for the year 2025. Although reported by the press in terms of miles per gallon, the standard is heavily written in terms of emissions limits to curb global warming.[28] Saving money on fuel costs can be good, but auto markets are being distorted by unfounded global warming fears.

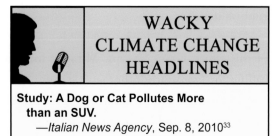

Fees on large vehicles are a government favorite to fight climate change. The owner of a subcompact car in Japan will pay about $4,000 less in taxes than the owner of a heavier passenger car during the life of the vehicle.[30] Great Britain established a carbon dioxide-based fee system for passenger vehicles back in 2001. Vehicle sales fees or annual registration fees in France, Germany, Italy, Spain, and most other European nations include taxes based on the size of emissions.

Ireland has introduced the world's most steeply graduated system of emissions-based auto taxes. An Irish owner of a high-emissions vehicle pays a 36-percent tax at sale compared to a 14-percent tax for a low-emissions vehicle. The same vehicle owner pays a €2,000 annual premium in registration taxes. The total emissions-based tax premium is about €19,000 ($25,270) over a five-year lifetime.[31] It's clear the Irish government doesn't want you to own that Land Rover®. Ireland annually emits 0.2 percent of global emissions, equal to the amount China emits in only 2.5 days. It's great that Irish drivers are making such a sacrifice for us all.

Hybrids and Plug-in Hybrid Electric Vehicles (PHEVs) are the rage. Hybrids, like the Toyota Prius®, are essentially gas-powered cars that use a battery and electric motor to improve gas mileage. Although misguided concerns about global warming are certainly a factor in hybrid sales, hybrid cars have closed much of the price gap with conventional cars, ten years after introduction. About 1.3 percent of global car sales in 2009 were hybrids.[32]

WACKY CLIMATE CHANGE HEADLINES

Study: A Dog or Cat Pollutes More than an SUV.
—*Italian News Agency*, Sep. 8, 2010[33]

PHEVs are true plug-in electric cars that use an auxiliary gasoline engine to extend driving range. Spurred on by climate mania, support for electric cars has become a mission for many of the world's leaders. President Obama pledged to "put a million plug-in hybrid cars…on the road by 2015."[34]

Governments of developed nations are competing to pay subsidies to manufacturers, as well as offer tax breaks of $5,000 to $8,000 to consumers who purchase a plug-in electric car. In 2010, the US provided a $151 million grant to a battery plant in Michigan.[35] The same year, the UK provided £20 million ($31 million) to Nissan for an electric car assembly plant in England.[36]

Yet, adoption of electric cars is likely to be disappointing. The first commercial PHEVs arrived on the market at the end of 2010 and start of 2011. They all feature short driving ranges and significantly higher prices than petroleum-powered cars. They all require at least two to three hours to recharge batteries. A 2010 Nielsen survey for *Financial Times* found that 65 percent of Americans and 76 percent of British citizens would not pay more for an electric car.[37] Consumers seem to have more sense than our political leaders. I wonder how many will pay more once they realize that driving an electric does nothing to improve the climate?

Your home—decarbonize it!
UK Prime Minster David Cameron states:

> …if we all turned down the thermostat in our house by just one degree, we would save over £650 million worth of energy and nearly nine million tonnes of carbon emissions every year. That would be the equivalent of taking three million cars off our roads…we can bring about a Green Consumer Revolution in this country to improve our lives, enrich our economy and protect our environment.[38]

It's a good bet that British homeowners already know they can save money by turning down their thermostats. But the idea that this can save polar bears should trigger one's rubbish detector.

Cameron has pledged to make his government the "greenest government ever." As part of this effort, Grant Shapps, Minister for Housing, announced a program in July 2007 calling for "zero-carbon homes," stating: "We will therefore ensure that from 2016 new homes need not add extra carbon to the atmosphere." What? How about heating, air conditioning, light bulbs, stoves, refrigerators, washing machines, water heaters, televisions, and computers? What about all the energy used to manufacture the materials, such as lumber, concrete, brick, plastic, and glass? Does this mean no more fireplaces or lamb steaks on the barbecue? *Even a grass hut in Cameroon exceeds zero carbon emissions.* Nevertheless, Minister Shapps hired the consultant Zero-Carbon Hub for £600,000 per year to establish standards to get to the "zero-carbon home."[39]

Even this Cameroon hut isn't "zero-carbon."

After five months of work, the Ministry issued a report that stepped back from the zero emissions goal but called for a 44-percent reduction from a 2006 home by 2013. The report recommends a "carbon compliance level" of less than 12 kg per year for each square meter of floor space for a house and under 14 kg/year/m^2 for low-rise apartments.[40] Since no one can actually measure the emissions from a television, an air conditioner, or most other parts of a house, computation of such carbon emissions will be loaded with guesswork. But it sure will create a big industry of regulators, inspectors, and consultants.

A key Climatist initiative for reducing homeowner emissions is switching light bulbs. Cuba banned the sale and importation of incandescent bulbs in favor of Compact Fluorescent Lights (CFLs) in 2005. Following the lead of this trend-setting country and urged on by Greenpeace, the European Union and all major developed nations have followed suit. Climate activists state that CFLs use less energy and reduce global greenhouse gas emissions, so incandescent bulbs must be banned.

Also claimed is that CFLs last longer, providing a cheaper alternative than traditional bulbs. But we do have that mercury problem with CFLs. The EPA's web site provides a three-page clean-up procedure if you happen to break a CFL, which stops just short of requiring a Hazmat suit.[41]

So, "How many environmentalists does it take to clean up a broken light bulb?" I live in Illinois, a state which outlaws disposal of CFLs in the regular trash pick-up. Bulb disposal requires a special trip in my carbon-emitting vehicle. I wonder if this was figured into the cost and

Climate Hypocrisy?

"We have pledged to be the greenest government ever. We must lead by example... We are not asking others to do things we will not do ourselves."
   —Charles Hendry, UK Minister of State for the Department of Energy and Climate Change (2010)

Climate Minister Buys a Castle with 16 Bathrooms...and a Massive Carbon Footprint
   —*Daily Mail*, Nov. 18, 2011[42]

CLIMATIST HIT LIST

The Incandescent Light Bulb[43]

emissions comparison with traditional bulbs?

The "smart meter" is another program to change your life in the name of protecting the environment. A smart meter is a device that goes in your house to monitor electricity or gas usage on an hourly basis. The idea is that consumers will be "rewarded" for using energy in off-peak periods. But smart meters may also have the capability to remotely shut off your electricity. This allows your government or utility to control your energy to avoid "wasteful usage" or reduce your "carbon footprint." Of course, consumers must pay to install the meters. Another great benefit is that employees at your electric utility will be able to tell when your family is at home. This breach in privacy has the potential to boost the residential burglary business.

In 2004, the government of Victoria, a province of Australia, mandated that its 2.6 million electricity customers should install smart meters. Andrew Blyth, CEO of the Australia Energy Networks Association, calls smart meters "an essential tool for tackling climate change." The forced installation of smart meters cost Victoria citizens and businesses $A2.25 billion and raised a homeowner's 2010 electric bill by $A110.[44]

The leaders of developed nations are now singing the praises of smart meters. The UK's Low Carbon Transition Plan of 2009 called for smart meters in every home by 2020.[46] In October 2009, President Obama announced 100 grants, totaling $3.4 billion, to begin work on a 20-year effort to create a smart grid in the US.[47] Certainly there are advantages in modernizing the electrical network of each nation. But the idea of rebuilding our networks to reduce our carbon footprint or enable more use of costly renewable energy is misguided.

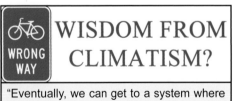

WISDOM FROM CLIMATISM?

"Eventually, we can get to a system where an electric company will be able to hold back some of the power so that maybe your air conditioner won't operate at its peak..."
—Carol Browner, former US Energy and Climate Change Advisor (2009)[45]

Your business—green it!

Global business has gone green—hook, line, and sinker. Today it's hard to find a major corporation that doesn't pitch the need to fight climate change. Sustainable environmental policies are favored by consumers, so companies adopt and promote green policies to improve their business image. Firms that are not proactive are often harassed by environmental groups. Much of the corporate climate control effort is aimed at reducing energy use. If based on sound economics, this is good practice, lowering cost and improving profits.

Leadership in Energy and Environmental Design (LEED) is a green building standard, developed by the US Green Building Council, that is increasingly used by US companies.[48] LEED promotes "sustainable" building practices in site selection, water usage, materials and other factors. LEED was meant to be a voluntary system, but US city and town councils are now making LEED methods mandatory for new construction to do their part for the climate.

But businesses today are making foolish decisions, based on the faulty science of man-made global warming. Many companies track carbon emissions for each of their processes. Companies such as Dow Chemical, General Electric (GE), Hewlett-Packard, IBM, Nike, PepsiCo, Toyota, and United Technologies have been lauded for their "best business practices" in energy efficiency and emissions control. In addition to their own efforts, these firms now require supplier companies to track greenhouse gas emissions.[49] Others are buying green energy from wind and solar fields at above-market prices, thereby unnecessarily increasing costs and reducing shareholder value. As we'll discuss in Chapters 4 and 5, science increasingly shows that global warming is due to natural processes, and man-made emissions play only a very minor role.

Some companies are profiting handsomely from climate mania. Climate-conscious products and services provide important sales revenue for these firms. GE now sells energy-efficient and environmentally-friendly products and services worth $20 billion per year.[50]

It's surprising that engineering and technology companies such as Hewlett-Packard, IBM, and Motorola accepted the theory of man-made global warming. These companies stand for excellence in engineering and technology. When

Is It Really about the Climate?

"Because the idea of climate change is so plastic, it can be deployed across many of our human projects and can serve many of our psychological, ethical and spiritual needs...We need to ask not what we can do for climate change, but to ask what climate change can do for us."
—Mike Hulme, Director of the Tyndall Centre for Climate Change Research and IPCC Lead Author (2009)[51]

such companies have a product field failure, they insist on rigorous root-cause analysis and experimentation to track down and resolve the problem. Yet, each has adopted the theory of man-made global warming, which has very little empirical basis.

US electrical utilities deserve a special mention, because they've sold out the American consumer. For years, these companies have struggled to raise prices to support capital investment in new generating facilities. State-level consumer groups have kept a lid on electricity rate increases. Many state governments now mandate renewable sources for electricity, such as wind and solar. Utilities know how poor wind and solar are as energy sources, but they've remained silent in the face of renewable mandates.

By adopting renewables, these companies gain an opportunity to raise electricity prices. In fact, they have supported cap-and-trade legislation in Congress, lobbying for grants of free emissions credits that will be worth hundreds of millions of dollars. These credits are intended to be passed through to the consumer in electricity rate savings, but some portion would instead go to company shareholders, and the consumer would see higher bills. Just another sad situation in a world of climate madness.

Your air travel—skip it!

Commercial air travel is a special problem for Climatism. There is no real low-carbon substitute for most of today's air travel. The only large-scale option for cutting emissions from aircraft is to encourage (force) people to fly less. Lord Turner, 2009 chairman of the UK Parliament committee on climate change, put it directly: "In absolute terms, we may have to look at restricting the number of flights people take."[52]

Britain has levied an eco-tax on air travel since 2007, called the Air Passenger Duty (APD). The APD has four rates based on trip distance, applying a tax of £24 ($37) per one-way flight for destinations to Europe and up to £170 ($263) for trips across the globe.[53] As of early 2011, Germany was the only other nation applying an air travel climate tax. Netherlands introduced a "departure tax" in 2008 which added 11–45 Euros to the cost of each flight. But after a steep decline in air traffic at Dutch airports, they rescinded the tax a year later. Travelers planning to fly through Netherlands switched to airports in other nations to avoid the tax.[54]

Since 2003, a battle has raged over adding a third runway to Heathrow Airport. British Airways and Her Majesty's Government have advocated the additional runway, opposed by local authorities, Greenpeace, and other environmental groups. Opponents purchased land in the path of the proposed expansion and engaged in sensational demonstrations, such as climbing on top of jets to unfurl banners. After eight years, efforts to construct the runway have stalled, hailed as a victory for the planet.[55]

Green Air Transportation?[57]

Yet, this is dwarfed by airport building in Asia. China is in the midst of a fourteen-year building program that will construct *47 new commercial airports* and expand 73 others.[56] India is building dozens of new airports and expanding more. One blocked runway at Heathrow is insignificant.

The 15,000 climate crusaders who flew to the 2010 Cancun Climate Conference didn't seem to be worried about their carbon footprint. Maybe that's because they bought "carbon offsets" at one of the kiosks located at world airports. Yes, you can now "offset" the carbon emissions of your plane travel with a payment toward planting trees in Indonesia or some other carbon-soaking project. According to the Green Flight Carbon Offset Calculator, an offset costs $65.90 for a round-trip flight from London to Cancun, which emits a single-passenger total of 3.66 tons of carbon.[58] This one trip emits *five times* the future per person *annual* emissions limit promised by world leaders, but who's counting?

High-speed rail is a recent nonsense solution for global warming in the United States. In April of 2009 the Obama administration announced a plan to build high-speed inter-city passenger rail in America, with a budget that eventually grew to $53 billion over six years.[59] A stated objective of the program was "promoting energy efficiency and environmental quality."[60] This program is really an effort to switch passengers from planes to trains to stop global warming. I'd like to take a leisurely vacation train ride from Chicago to the West Coast someday. But why does any rational person want to pay more and take longer to travel between cities that are already linked by today's commercial air flights?

Air travel and global warming fears started the first climate change trade war in early 2012. As of January 1, 2012, the European Union (EU) required airlines flying to Europe to join the European emissions trading scheme and pay a carbon tax.[61] Twenty-one countries, including Brazil, China, India, Japan, Russia, and the United States have issued declarations opposing the EU mandate.[62] The government of China forced Hong Kong Airlines to delay purchase of a multi-billion-euro aircraft order with European manufacturer Airbus.[63] Undaunted, European climate action commissioner Connie Hedegaard

PETA encourages Al Gore to go vegetarian.[64]

stated, "There is no way the EU will change legislation [on aircraft emissions]" and that government attempts to forbid participation by airlines were "arrogant and ignorant."[65]

Your diet—veg it!

The organization People for the Ethical Treatment of Animals (PETA) declares:

> The evidence is in…UN report shows that animals raised for food generate more greenhouse gases than all cars and trucks combined.[66]

If you accept the theory of man-made global warming, PETA is correct. Vegetarian Rajendra Pachauri, Chairman of the IPCC, lectures us:

> Meat production represents 18 per cent of global human-induced GHG emissions…While the world is looking for sharp reductions in greenhouse gases responsible for climate change, growing global meat production is going to severely compromise future efforts…a study from the University of Chicago showed that if Americans were to reduce meat consumption by 20 per cent it would be as if they switched from a standard sedan to the ultra-efficient Prius.[67]

Indeed, not only agricultural processes emit greenhouse gases, but cattle and hogs literally belch out large quantities of methane, a greenhouse gas. In 2009, the British National Health Service (NHS) issued the report *Saving Carbon, Improving Health*, which calls for:

> The actions needed to develop a more sustainable food system in the NHS whilst maintaining nutritional value include…a reduction in the reliance on meat, dairy and eggs.[68]

Musician and vegetarian Paul McCartney has called on the world to adopt "Meat-Free Monday."[69] The idea is to forego meat at least one day per week to fight climate change. Supporters include Alec Baldwin, Sheryl Crow, and Richard Branson, who says:

> I love eating meat, but I love our planet even more. So I will join this campaign and stop eating meat at least one day each week.[70]

On Sir Paul's advice, the San Francisco city council named Monday "Vegetarian Day." The town of Ghent, Belgium adopted Thursday as a meat-free day.[71] For centuries, Catholicism required Friday to be a meat-free day. It seems that this has now been replaced by another day for believers in Climatism.

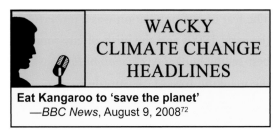

WACKY CLIMATE CHANGE HEADLINES

Eat Kangaroo to 'save the planet'
—*BBC News*, August 9, 2008[72]

If you're someone who just can't give up meat, there is an alternative. The United Nations is now considering to recommend greater consumption of insects as a way to limit global warming. Professor Arnold van Huis, an entomologist at the Wageningen University in Belgium, found that if people eat bugs instead of meat, environmental impacts will be greatly reduced. According to van Huis, breeding of *meal worms, crickets, and locusts* emits 10 times less methane than livestock, and 300 times less nitrous oxide than pig and poultry farming.[73] Bon appetit!

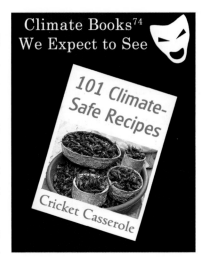

Climate Books[74] We Expect to See

101 Climate-Safe Recipes

Cricket Casserole

Climatists also tell us that fat people are causing more climate damage than thin people. Sir Jonathan Porritt, Chairman of the UK Sustainable Development Commission, pointed out in 2009 that overweight people eat more protein-rich food such as beef or lamb, which comes from livestock that emit the greenhouse gas methane. He also said obese people are more likely to use cars rather than walk or cycle, thereby producing more carbon emissions. Porritt states:

The World Health Organization recently published some data showing that each overweight person causes an additional one tonne of CO_2 to be emitted every year. With one billion people overweight around the world—of whom at least 300 million are obese—that's an additional one billion tonnes.[75]

WACKY CLIMATE CHANGE HEADLINES

Researchers Suggest Link Between Obesity & Global Warming
—*Scotland of Food and Drink*, August 17, 2011[76]

Your family—downsize it!

The fear of overpopulation was a powerful foundation for the conservation movement in the early 1900s and the birth of environmentalism in the 1960s. Dr. Paul Ehrlich's 1968 best-selling book *The Population Bomb* warned that Earth's growing population was pushing mankind toward catastrophe:

> The battle to feed all of humanity is over. In the 1970s the world will undergo famines— hundreds of millions of people will starve to death in spite of any crash programs embarked upon now.[77]

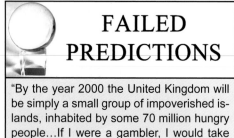

FAILED PREDICTIONS

"By the year 2000 the United Kingdom will be simply a small group of impoverished islands, inhabited by some 70 million hungry people...If I were a gambler, I would take even money that England will not exist in the year 2000."
—Dr. Paul Ehrlich (1971)[78]

Contrary to alarming predictions, population growth slowed dramatically over the last 50 years. Global population growth peaked at about two percent per year in 1970, but dropped to just above one percent per year by 2007. According to United Nations data, 74 nations now have birth rates below 2.1 children per woman, considered the zero population growth rate, including Australia, Brazil, Canada, China, all European nations, Japan, Russia, South Korea, and the United States.[79] Even though fertility rates are below zero growth in these nations, birth rates continue to decline. Fertility rates in developing nations are now less than 2.5 children per woman, down from 5.3 per woman in 1970.[80] Birth rates are declining in almost all nations of the world.

Even though Dr. Ehrlich was spectacularly wrong, overpopulation has become a foundation of Climatism. The idea is simple. Climate alarmists point out that every human activity uses energy, and energy use emits greenhouse gases, therefore population control is essential to stop global warming. According to Chris Rapley, director of the British Antarctic Survey:

> Although reducing human emissions to the atmosphere is undoubtedly of critical importance...the truth is that the contribution of each individual cannot be reduced to zero. Only the lack of the individual can bring it down to nothing.[81]

The UN makes it clear that people are the problem:

> The importance of the speed and magnitude of recent population growth in boosting future greenhouse gas emissions is well recognized among scientists...Each birth results not only in

the emissions attributable to that person in his or her lifetime, but also the emissions of all his or her descendents. Hence, the emissions savings from intended or planned birth multiply with time…No human is genuinely "carbon neutral," especially when all greenhouse gases are figured into the equation. Therefore, everyone is part of the problem, so everyone must be part of the solution in some way.[82]

WISDOM FROM CLIMATISM?

"We're too many people; that's why we have global warming…on a voluntary basis, everybody in the world's got to pledge to themselves that one or two children is it."
—Ted Turner, media mogul and father of five children (2008)[83]

Just think of all the "emissions savings" we can amass if we just "lack individuals," since no one can be "carbon neutral."

China established a one-child policy in 1978. Over the last 40 years, this policy and other demographic factors have resulted in a reduction in the population growth in China. But it's widely reported that the program has also been responsible for infanticide, forced abortion, forced sterilization, abandonment of infants, and sex-selective abortion using ultrasound testing. Zhao Baige, Vice Minister of China's National Population and Family Planning Commission, pointed out the climate benefits at the 2009 Copenhagen Climate Conference:

> The policy on family planning proves to be a great success. It not only contributes to reduction of global emission, but also provides experiences for other countries…in their pursuit for a coordinated and sustainable development.

Zhao went on to estimate that China had 400 million fewer births and emitted 1.8 billion fewer tons of carbon each year because of the program.[84]

Your government and economy—change it!

The growth of democracy, the free-market economy, and advances in technology have been the foundation of human prosperity over the last 100 years. Between 1900 and 2005, human life expectancy more than doubled from 30 years to 70 years. Personal annual income increased from about $1,000 to more than $6,000 during the same period. According to Dr. Bjorn Lomborg of the University of Aarhus in Denmark:

WISDOM FROM CLIMATISM?

"…the single-most concrete and substantive thing an American, young American, could do to lower our carbon footprint is not turning off the lights or driving a Prius, it's having fewer kids…"
—Andrew Revkin, Environmental Reporter
New York Times (2009)[85]

…children born today—in both the industrialized world and the developing countries—will live longer and be healthier, they will get more food, a better education, a higher standard of living, more leisure time and far more possibilities—without the global environment being destroyed.[86]

Despite the advances of mankind in the last century, the specter of catastrophic climate change is now used to call for elimination of democracy. David Shearman and Joseph Smith argue in *Climate Change and the Failure of Democracy* that our democratic form of government will not be able to cope with climate change.[87] They advocate replacement of democracy with autocratic forms of government to solve the crisis. Dr. James Lovelock, English scientist and environmentalist agreed:

We need a more authoritative world…What's the alternative to democracy? There isn't one. But even the best democracies agree that when a major war approaches, democracy must be put on hold for the time being. I have a feeling that climate change may be an issue as severe as a war. It may be necessary to put democracy on hold for a while.[88]

Capitalism is also under attack. Some view capitalism and economic growth as unchecked forces that will destroy the climate and the planet. British author Robert Newman states:

Capitalism is not sustainable by its very nature. It is predicated on infinitely expanding markets, faster consumption and bigger production in a finite planet…You can either have capitalism or a habitable planet. One or the other, not both.[89]

Major leftist leaders of the world fault capitalism for climate change, including Evo Morales of Bolivia, Mahmoud Ahmadinejad of Iran, and Hugo Chavez of Venezuela. According to President Morales:

"If we don't overthrow capitalism, we don't have a chance of saving the world ecologically."
—Judi Bari, environmental group Earth First[90]

Capitalism and the thirst for profit without limits of the capitalist system are destroying the planet…Climate change has placed all humankind before a great choice: to continue in the ways of capitalism and death, or to start down the path of harmony with nature and respect for life.[91]

The UN indirectly attacks capitalism, by calling for "sustainable patterns of production and consumption." Today's website of the United Nations Environment Programme (UNEP) laments the spread of consumerism:

These western lifestyles of consumerism are spreading all around the world through products and services, media and trade policies. Western type restaurants and coffee shops are as common on the streets of Beijing, as international brands of clothing and other products...Goods and services previously seen as luxuries—TVs, mobile phones and cars— have now become necessities. The supply of goods from exotic locations is increasing, as well as the consumption of processed food and meat.[92]

Is It Really about the Climate?

"We've got to ride this global warming is-sue. Even if the theory of global warming is wrong, to have approached global warming as if it is real means energy conservation, so we will be doing the right thing anyway in terms of economic and environmental policy."
—Timothy Wirth, former US Senator, President of the UN Foundation (1992)[93]

So what must be done? According to Professor Kevin Anderson, Director of the Tyndall Centre for Climate Change Research in the United Kingdom, we must halt the economic growth of the developed nations for the next 20 years. He advises:

> The Second World War and the concept of rationing is something we need to seriously consider if we are to address the scale of the problem we face.[94]

Former UK Deputy Prime Minister John Prescott agrees: "What we are beginning to witness is a whole new set of rules for economics, based on rationing resources."[95]

So, the message of Climatism is clear. In order to save the planet, we must adopt sacrificial policies that will reduce the freedom of every citizen. We must forgo SUVs and pickup trucks and switch to electric cars. We must eliminate air travel vacations and visits to distant relatives. We must change our light bulbs and "green" our businesses. Vegetarianism and small families are in and large pets are out, so we can reduce our carbon footprint. Democracy must be replaced with an autocratic government that can lead us toward environmental paradise. Capitalism, overproduction, and wasteful consumerism must be replaced with "new rules for economics, based on rationing resources." Join the climate revolution.

THE SPECTER OF GLOBAL GOVERNANCE

Since democratic governments are not up to the task, climate alarmists advocate establishment of "global governance" to solve the climate crisis. The United Nations, as the leading organization for Climatism, is ready to step up to the plate as that global governor. In fact, the UN has been using the issue of global warming to build its position of global governance for the last 25 years.[96]

The UN established the extra-national IPCC in 1988 to drive the vision of man-made global warming. The UN negotiated the Framework Convention on Climate Change (FCCC) in 1992, an international treaty signed by 41 nations and the European Community, to reduce greenhouse gas emissions. The signing of 1997 Kyoto Protocol was the next step along the path. The UN-sponsored Protocol was eventually ratified by 184 nations (the US was the major exception), binding industrialized nations to hard numbers for emissions reduction.

Jacques Chirac, former President of France, praised the 1997 Kyoto Protocol climate treaty as the "first component of an authentic global governance."[98] Prior to the 2009 Copenhagen climate conference, UN Secretary General Ban Ki-moon called for an "equitable global governance structure" for controlling carbon emissions.[99] The climate conferences in Copenhagen in 2009, Cancun in 2010, and Durban, South Africa, in 2011 have been efforts to establish a new international treaty to replace the Kyoto Protocol. Expect another climate conference every year in the name of reducing greenhouse gases, but designed to move toward a greater degree of centralized global control.

The UN views national sovereignty as the primary impediment to a UN-led system of global government, although this view is not often publicly expressed. Maurice Strong, nicknamed the "father of the Kyoto Protocol," divulged this UN position while Executive Director of the United Nations Environment Programme (UNEP) at the 1992 Earth Summit:

The concept of national sovereignty has been

JOIN THE CLIMATE REVOLUTION!

"Climate change is also showing us that the old model is more than obsolete. It has rendered it extremely dangerous. Over time, that model is a recipe for national disaster. It is a global suicide pact...it may sound strange to speak of revolution. But that is what we need at this time."
—Ban Ki-moon,
    UN Secretary General (2011)[97]

THE GORE EFFECT

Blizzard Dumps Snow on Copenhagen as Leaders Battle Warming
"World leaders flying into Copenhagen today to discuss a solution to global warming will first face freezing weather as a blizzard dumped 10 centimeters (4 inches) of snow on the Danish capital overnight...Denmark...hasn't had a white Christmas for 14 years...and only had seven last century."
—*Bloomberg*, Dec. 17, 2009[100]

an immutable, indeed sacred, principle of international relations. It is a principle which will yield only slowly and reluctantly to the new imperatives of global environmental cooperation. It is simply not feasible for sovereignty to be exercised unilaterally by individual nation states, however powerful. The global community must be assured of environmental security.[101]

Part of the UN plan to convince nations to give up their national hegemony, is to establish a World Environmental Organization (WEO). At the annual meeting of the United Nations Environment Program (UNEP) in February, 2010, a high-level ministerial group was established to discuss the formation of a WEO. UNEP's Executive Director, Achim Steiner, told reporters:

> The status quo…is no longer an option. Within the broader reform options, the WEO concept is one of them.[102]

If established, the WEO would have the power to impose sanctions on carbon-polluting or over-consuming nations to protect the environment.

The IPCC, WEO, and a successor climate treaty to the Kyoto Protocol are key elements of the UN effort to use the issue of climate change to establish world governance. Once established, the UN will use these organizations to reduce "overpopulation," "over-consumption," and "overproduction." By controlling carbon emissions, the UN seeks to control global economic development and retard the growth of developed nations, especially the United States. Vaclav Klaus, President of the Czech Republic, warns:

> I am afraid there are people who want to stop the economic growth, the rise in the standard of living (though not their own) and the ability of man to use the expanding wealth, science and technology for solving the actual pressing problems of mankind, especially of the developing countries.[103]

Second, the UN seeks to redistribute wealth from the developed nations to the developing nations. Attendees to the 2010 Cancun Conference agreed to establish a "Green Climate Fund," a proposal reinforced at the 2011 Durban Conference. Industrialized nations are to provide $30 billion by 2012, and rising to an annual $100 billion by 2020.[104] The funds are to be administered by the World Bank and

Is It Really about the Climate?

"But one must say clearly that we redistribute de facto the world's wealth by climate policy…One has to free oneself from the illusion that international climate policy is environmental policy any more."
—Ottmar Edenhofer, Co-chair of IPCC Working Group III, Lead Author (2010)[105]

distributed to poor nations to pay for the impacts of climate change and to assist with low-carbon development. Of course, climate alarmists claim this effort is morally justified because the wealthy nations have created the climate crisis with decades of greenhouse gas emissions.

THE DILEMMA OF DEVELOPING NATIONS

Energy is the lifeblood of prosperity. The high-income developed nations of the world use a high level of energy per person. Poor nations suffer from low energy usage and low penetration of vehicles.

For example, the average per-person income for a Swede is about $33,000 per year, about 30 times the average income of an Indonesian, at about $1,000 per year. Sweden has one car for every two citizens, while only about one in 20 persons in Indonesia has a car. Sweden uses about 15,238 kilowatt-hours (kW-hr) of electricity per person, one of the highest usages in the world, compared to 564 kW-hr per person in Indonesia, which is among the lowest. Of the 80 million people in Indonesia, almost half do not have access to electricity. On average, each person in Sweden uses seven times the energy of an Indonesian.[106] During the last century, western economies were built on high energy usage, electricity, and hydrocarbon vehicles.

Despite this reality, Climatism seeks to deny the advantages of hydrocarbon energy to developing nations. According the United Nations 2010 report "Low Carbon Development Path for Asia and the Pacific," the developing nations must change their ways:

> Asia-Pacific countries must undergo structural adjustment to make key policy changes needed to switch their development mode…Most member countries have followed the industrial model of developed countries, which is the root cause of climate change. This traditional industrial development model results in an unsustainable energy consumption pattern.[107]

The report discusses how Asia Pacific must "pursue a low carbon development path" and skip a "growth path heavily reliant on pollutants." The authors recognize that Asian nations need to continue to pursue economic development, but that "luxurious or wasteful emissions (viewed as those that do not meet basic human needs such as shelter or food) should be discouraged." *The report goes on to ask whether televisions, computers, and social networking through the internet are necessary activities.*[108] UN officials seem to have no remorse about demanding such changes from poor nations, while flying off to Bali, Cancun, or other exotic locations every year to attend global climate conferences.

The poor people of the world need hydrocarbon energy to boost their standard of living, reduce disease, and extend life spans. One serious problem for millions of people in poverty is caused by use of primitive energy sources. Women in Africa walk miles each day to gather wood, charcoal, and dung, used for heating and cooking. These materials are burned inside huts, causing harmful indoor air pollution.

According to the World Health Organization:

> Every year, indoor air pollution is responsible for the death of 1.6 million people—that's one death every 20 seconds...In sub-Saharan Africa, the reliance on biomass fuels appears to be growing as a result of population growth and the unavailability of, or increases in the price of, alternatives such as kerosene and liquid petroleum gas.[109]

An "energy pipeline" in Ivory Coast, Africa
(Photograph by Zenman)[110]

Hydrocarbon fuels are the answer to eliminate indoor air pollution and to halt the practice of cutting down forests for fuel. Yet, Climatists recommend that poor nations forgo use of hydrocarbon energy. Instead they advocate the use of "small hydro," wind, biofuels, solar power, and electric cars, all of which are expensive and entirely inadequate for the size of the energy needs of poor nations.

In April, 2010, the World Bank approved a $3.9 billion loan to the nation of South Africa for the planned Medupi power station, to be the fourth largest coal-fired power station in the world. But this loan was only approved because the developing nation representatives of the World Bank board voted for approval. The US member abstained from approval because of:

> ...concerns about the climate impact of the project and its compatibility with the World Bank's commitment to be a leader in climate change mitigation and adaptation.[111]

The UK and three other European nations joined the US in disapproval. It's the sorry position of our governments that Africans should

WISDOM FROM CLIMATISM?

🚲 WRONG WAY

"To build a power plant and run lines to houses, to huts, to anything is a tremendous amount of work...how about...just giving them the service where they need it—on the roof of their hut."
—Environmentalist Ed Begley, Jr. (2007)[112]

IS THIS WHAT YOU MEAN BY SUSTAINABLE???

Tehachapi Wind Farm, California[113]

not have electricity for refrigerators to reduce food-spoilage disease, for hospitals to improve health, and for factories to build economic growth and reduce poverty if such electricity is produced by hydrocarbon power plants.

Although the Medupi loan was approved, many loans are rejected. Environmental groups such as BankTrack, Friends of the Earth, and Rainforest Action Network have forced most major banks to sign the "Equator Principles." The principles demand that banks lend only in an "environmentally responsible" manner. This responsibility increasingly precludes lending to projects involving oil, gas, and coal-fired power plants and production facilities, along with large dams and nuclear plants. Under tremendous pressure, Citibank, J.P. Morgan Chase, Bank of America, and most other banks of the world have surrendered and signed the Equator Principles.[114]

Because of the misguided ideology of Climatism, the growth of hydrocarbon energy will be limited and millions will continue to suffer in the developing world—a form of eco-genocide. Paul Driessen, author of the book *Eco-Imperialism*, summarizes:

> To block the construction of centralized power projects as not being "appropriate" or "sustainable" is to condemn billions of people to continued poverty and disease—and millions to premature death.[115]

THE CAP-AND-TRADE CARBON TRADING MESS

According to many, the solution to reach a low-carbon world is to raise the cost of hydrocarbons by "putting a price on carbon," thereby making renewable fuels comparatively more attractive. Australia Prime Minister Julia Gillard is an advocate:

> If we change the way the electricity sector operates, we can bring down our levels of carbon pollution, and continue the crucial task of tackling climate change. Putting a price on carbon would do this.[116]

Enter the "cap-and-trade" carbon trading system. Cap-and-trade is a system designed to place an indirect tax on carbon dioxide emissions, yet attempt to use market forces to encourage emissions reductions. Government authorities set a total emissions limit (the cap) for each participating industry. Emissions allowances or permits, which are rights to emit, are then sold or issued each year to participating firms. Companies may buy or sell allowances (the trade), but must submit allowances to the government each year according to the actual carbon dioxide emissions by their business. By

How big is the carbon footprint of *your* pet?[117]

issuing fewer permits each year, authorities intend to reduce total emissions over time. The system boosts the cost of generating electricity from coal, petroleum refining, and other industrial processes that are heavy users of hydrocarbon energy.

Cap-and-trade is an effort to create a purely artificial market to force people to change their behavior. Unlike a bushel of corn, a tangible commodity that people can buy and sell, no one has the foggiest idea about a ton of CO_2. Truly, it's a system for trading hot air.

Why do political leaders like cap-and-trade so much? The answer can be summarized in one word—*control*. In a cap-and-trade system, the entire economy is directed by an endless series of arbitrary decisions made by government bureaucrats. Regulators decide which industries are to be covered and which are exempt. They decide the level of emissions each year. They decide how many free allowances are to be issued and how many allowances are to be sold or auctioned. They decide which industries and ultimately which companies will get free permits worth billions of dollars. Bureaucrats estimate emissions for every industrial activity, and determine the "best emissions practice" or "best available technology." Want to punish the oil companies? Cap-and-trade is a great tool. Want to boost the wind turbine business? Cap-and-trade is just the ticket. As Dr. Richard Lindzen, Professor of Meteorology at the Massachusetts Institute of Technology, has observed:

> Controlling carbon is a bureaucrat's dream. If you control carbon, you control life.[118]

The Emissions Trading System (ETS), which began operation in 2005, is the world's largest cap-and-trade system. The ETS is administered by the European Community in cooperation with 30 participating nations and imposes emissions requirements on over 10,000 firms. Today's global emissions trading market totals €140 billion ($175 billion),[119] with most trading activity occurring in Europe in the ETS.

By any objective standard, the ETS has been a failure to date. During Phase I (2005-2007), participating nations set emissions caps higher than actual emissions. Allowances were then freely provided to participating firms. But too many allowances were issued, causing the price of carbon to drop from €33 per ton to zero by September 2007. European emissions *increased* by 1.9 percent from 2005 to 2007.[120]

ETS Phase II (2008-2012) is in process. Allowances are still provided for free, but fewer in number. But Phase II introduced the use of Clean Development Mechanism (CDM) credits that were created under the Kyoto Protocol.[121]

CDM credits are issued by the United Nations. Companies can purchase the credits to "participate" in emissions reductions projects in developing nations. These credits can then be submitted to governments in place of allowances required in ETS, allowing European companies to buy credits instead of reducing emissions. CDM projects act effectively as wealth transfers from the industrialized nations to the developing nations.

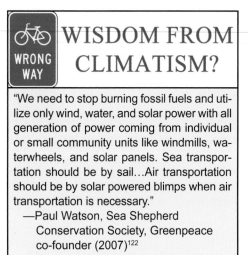

WISDOM FROM CLIMATISM?

"We need to stop burning fossil fuels and utilize only wind, water, and solar power with all generation of power coming from individual or small community units like windmills, waterwheels, and solar panels. Sea transportation should be by sail...Air transportation should be by solar powered blimps when air transportation is necessary."
—Paul Watson, Sea Shepherd
    Conservation Society, Greenpeace
    co-founder (2007)[122]

At the Carbon Expo trade fair held every year in Barcelona, nations from Africa and Asia offer climate-aid projects for investment by European firms. RWE, the giant German utility, has a 32-member team for identifying CDM projects around the world for investment.[123] The world market for CDM credits was worth $2.7 billion in 2010.[124]

CDM credits are defined to be only for emissions-reduction projects that would not be otherwise built, but smart project managers claim credits for projects that would be built anyway. Xiaogushan Dam in China was awarded $30 million in credits, even though construction was already underway and loans had been granted by the Asian Development Bank.

"Exotic" industrial gases have become a large share of total CDM credits. These are chlorofluorocarbons such as HFC-23, which are claimed to be potent greenhouse gases. Chemical makers are paid as much as $100,000 to eliminate a ton of HFC-23, so the CDM system actually encourages manufacturers to produce HFC-23, so that it can be eliminated. Mark Roberts of the research group Environmental Investigation Agency observes:

The evidence is overwhelming that manufacturers are creating excess HFC-23 simply to destroy it and earn carbon credits.[125]

The ETS is an excellent platform for scandal. Both organized crime and common crooks are making money in carbon credits the old-fashioned way—through money laundering and outright theft. The Danish CO_2 Quota Register became the largest carbon trading market in the world, primarily due to ease of registration. By the end of 2009 when the scandal broke, 1,256 carbon permit traders were listed on the Register. More than 80 percent of these brokers were front companies for money laundering and tax fraud schemes, costing the treasuries of Denmark and other European nations €5 billion ($7 billion). After governments modified their tax codes to remove carbon trading loopholes, trading volumes dropped by 90 percent.[127]

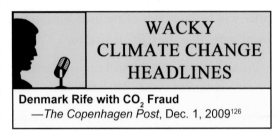

WACKY CLIMATE CHANGE HEADLINES

Denmark Rife with CO_2 Fraud
—*The Copenhagen Post*, Dec. 1, 2009[126]

In January 2011, the European Union closed all carbon credit spot markets in France, Germany, the Netherlands, Slovakia, and the United Kingdom for a period of two weeks. Hackers had stolen €30 million in carbon credits. This was the fourth time that European carbon markets were hit with a major scandal.[128]

But the worst of the ETS, the impact on European businesses and citizens, is yet to come. ETS Phase III will begin in 2013, when all emissions permits must be purchased. For example, RWE emits about 149 million tons of CO_2 per year to produce electrical power, more than any other European firm. Permits will cost the company over €2 billion per year. Most of this cost will be passed on to German citizens in the form of higher electricity rates.

Europe cement firms will be able to emit only 766 grams of CO_2 for each kilogram (kg) of cement produced. Other limits are 1,328 grams/kg for steel production and 1,514 grams/kg for aluminum production. Companies that exceed these limits will be required to purchase allowances.[129]

Depending upon the type of product produced, these limits can be onerous. In all cases, companies must track emissions for every product produced, which is not a simple process. The paper industry produces about 3,000 products, but the European Commission lists only 53 products contained in 7 categories.[130] Companies must deal with a blizzard of paperwork and decisions about which product categories and laws apply. The

ETS creates huge competitive imbalances within Europe, along with large disadvantages relative to other global companies that don't have such carbon regulations.

But some say that even cap-and-trade isn't enough. Rather than let the carbon trading market determine the price of carbon, the United Kingdom established its Carbon Reduction Commitment (CRC) scheme in 2010. The scheme sets a minimum price on carbon dioxide of £12 per ton and forces 5,000 businesses to pay for each ton of CO_2 they emit. The floor price is expected to rise to £30 per ton by 2030. Business costs from the carbon price will be hundreds of thousands to millions of pounds per year. The British government states that the CRC is vital to:

> …achieving the UK's overall targets of reducing greenhouse gas emissions by 2050 by at least 80% compared to the 1990 baseline.[131]

British businesses are worried that the tax will not allow them to operate. Tata Steel, the largest private business in Wales, is one of many businesses concerned with the impact of the CRC. Karl-Ulrich Kohler, head of Tata's European operations, labelled the measure "a potentially severe blow to the sustainability of UK steelmaking."[132]

Australia has gone even farther, becoming the first major nation to approve a direct tax on carbon dioxide. Beginning in July of 2012, 500 of Australia's largest companies will pay $A24 per ton of carbon dioxide emitted.[134] The regulations will place a large burden on energy and heavy industrial firms of the nation. The tax was approved by the government despite polls that showed that the majority of Australians were in opposition. Cap-and-trade and carbon taxes are monsters spawned from the misguided theory of man-made global warming. These policies are inflicting real and sizable damage on citizens and businesses across the world. Along with wind turbine towers and carbon capture pipelines, they're a testimony to the madness of Climatism.

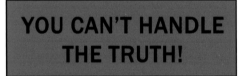

YOU CAN'T HANDLE THE TRUTH!

Australians to Punish Carbon Tax Criticism
"…the Australian Competition and Consumer Commission, which this week issued warnings to businesses that they will face whopping fines of up to $1.1m if they blame the carbon tax for price rises."
—*US Action News*, Nov. 17, 2011[133]

THE MAD, MAD, MAD WORLD OF CLIMATISM

Halting global warming is the top priority of the ideology of Climatism. It's more important than economic development, freedom, democracy, capitalism, and the welfare of poor nations. It's certainly more important than your car, your family, your diet, your home, your travel, and your lifestyle.

Welcome to the mad, mad, mad world of Climatism. It's a world of veggie meals and insect snacks, tiny cars, fluorescent bulbs, and carbon credits. It's a world where your children are measured by the size of their carbon footprint, rather than the content of their character. It's a world directed by global bureaucrats who are not accountable to any voter. It's a world of government-subsidized and expensive wind, solar, and biofuel energy. The proponents of Climatism assert that all this, and more, is necessary if we are to save the planet.

To accept this hardship, powerful science must point to the coming climate disaster if we don't change our ways. Let's examine the scientific basis for the theory of man-made global warming in the next chapter.

CHAPTER 3

THE SIMPLE SCIENCE OF MAN-MADE GLOBAL WARMING

"For every complex problem there's a simple answer—that's wrong!"
—AMERICAN JOURNALIST HENRY LOUIS MENCKEN

The theory of man-made global warming is simple and can be expressed in one sentence: "Man-made emissions of greenhouse gases are warming the planet dangerously." This is the "simple answer" used to explain the increase in Earth's surface temperatures in the late twentieth century. From this answer, the ideology of Climatism has plunged mankind into a madness affecting every person on Earth. Here we'll summarize the four basic premises that provide the foundation for the theory of man-made global warming. In Chapter 4 and Chapter 5 we'll show just how shaky this foundation is.

47

BASIS ONE: RISING GLOBAL SURFACE TEMPERATURES

The first basis for the theory of man-made global warming is that Earth's surface temperature is rising. Global surface temperatures are estimated to have increased about 1.3°F (0.7°C) since the late 1800s. We'll refer to this recent rise in temperatures as the "Modern Warming" when we discuss it in Chapter 4.

The Climatic Research Unit (CRU) of East Anglia University, in cooperation with the Hadley Centre of the UK Meteorological Office, maintains the leading data set for global temperatures.[1] Temperatures are tracked monthly from over 3,000 land stations around the world. Sea surface temperature data is reported from volunteer merchant and naval vessels. The data is adjusted, assembled and expressed as a "temperature anomaly," which is a difference from the 1961–1990 temperature average.

A rise of about one degree over the last 130 years doesn't seem like much. In Chicago, daily temperatures swing about 100°F in a single normal year. But proponents of man-made warming claim that this temperature rise is *unprecedented* in modern history. The IPCC stated in its Third Assessment Report of 2001:

> …the increase in temperature in the 20th century is likely to be the largest of any century during the past 1,000 years.[2]

If one concludes that the Modern Warming period of the last century is abnormal, then something abnormal must be causing it.

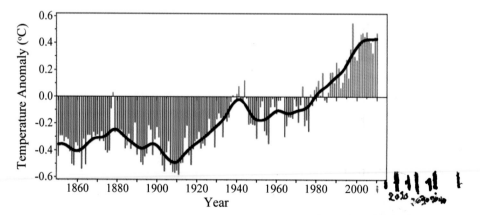

Global Surface Temperatures 1850–2010. Averaged global surface air temperatures from land and surface measuring stations. Temperature anomaly shows difference from the 1961–1990 average. Data from the Climatic Research Unit and the UK Meteorological Office Hadley Centre. (HadCRUT3, 2011)[3]

BASIS TWO: RISING ATMOSPHERIC CARBON DIOXIDE

The second basis for the theory of man-made global warming is rising atmospheric carbon dioxide concentration. Modern measurements of CO_2 levels began in 1958 at the Mauna Loa Observatory on the island of Hawaii. Dr. Charles Keeling, a young researcher under the direction of Dr. Roger Revelle, first measured atmospheric CO_2 at 315 parts per million (ppm). Called the "Keeling Curve," the data now show the CO_2 level to exceed 390 ppm. Other scientists estimate that the pre-industrial "background level" of atmospheric CO_2 was about 280 ppm before 1900.

The proponents of man-made global warming claim that this rise is primarily due to man-made emissions of carbon dioxide. The IPCC warns that CO_2 emissions have "perturbed the Earth's carbon cycle," causing global climate change. This curve is also presumed to be an abnormal climatic situation, supporting charges that carbon dioxide emissions from industry are pollution that must be controlled and eliminated.

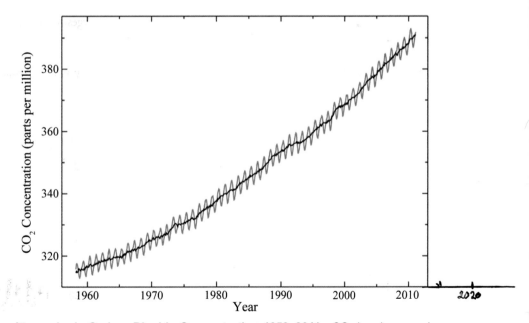

Atmospheric Carbon Dioxide Concentration 1958–2011. CO_2 has increased from about 315 ppm in 1958 to 392 ppm today. The red sawtooth curve shows the seasonal CO_2 variation while the black line is the average. All atmospheric gas concentrations in this book are in parts per million by volume (ppmv). Data from the Earth System Research Laboratory in Mauna Loa, Hawaii. (NOAA, 2011)[4]

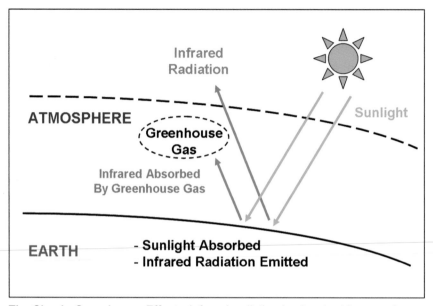

The Simple Greenhouse Effect. Infrared radiation is absorbed by greenhouse gases in the atmosphere.

BASIS THREE: THE GREENHOUSE EFFECT

The third basis for the theory of man-made global warming is the greenhouse effect. Sunlight, which is high-energy radiation, enters our atmosphere and is absorbed by Earth's surface. Like any warm body, Earth emits radiation. Since the Earth's temperature is lower than that of the sun, Earth emits lower-energy radiation called infrared radiation, which is not visible to our eyes. Some of the infrared radiation passes back out of our atmosphere into space, but most is absorbed by greenhouse gases in our atmosphere. The warming caused by this absorption of infrared radiation is called the greenhouse effect.

Greenhouse gases are those which strongly absorb infrared radiation. Water vapor and carbon dioxide are Earth's most important greenhouse gases, but methane, nitrous oxide, ozone, chlorofluorocarbons, and other gases are also greenhouse gases. After absorbing outgoing infrared radiation, these gases reradiate a portion of the captured energy back to Earth. This acts to warm Earth's surface.

The greenhouse effect is the theoretical basis for the theory of man-made global warming. Most greenhouse gases in Earth's atmosphere are created by natural climatic processes, so the greenhouse effect is a natural effect. But emission of greenhouse gases from human industry adds to the effect. Mr. Gore and others raise the alarm that man-made emissions are the cause of both rising atmospheric CO_2 concentration and rising global surface temperatures, which they fear may prove dangerous. *But quantifying the amount of the greenhouse effect contributed by mankind is key to estimating the impact.* However, the human contribution has not been quantified by climate alarmists.

BASIS FOUR: COMPUTER MODEL PROJECTIONS

The fourth basis for the theory of man-made global warming is computer model projections. For the last 40 years, General Circulation Models (GCMs) of increasing complexity have been used to model Earth's climate. GCMs use initial starting data, the laws of physics and thermodynamics, and lots of computing power to simulate Earth's climate. The models are tuned from past climate history and then run over and over to forecast the climate far into the future. Model outputs include temperature, air speed and direction,

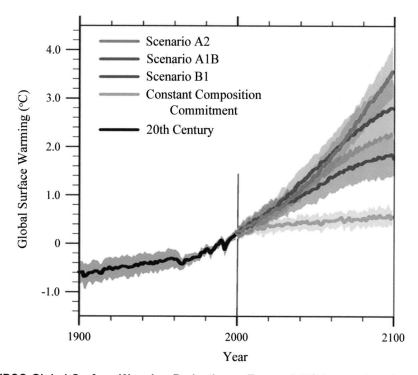

IPCC Global Surface Warming Projections. From a 0.7°C temperature rise during the last 130 years, an averaging of computer model projections predicts a rise of 3°C (5.4°F) in the next 90 years. (IPCC AR4, 2007)[5]

air pressure, and humidity at thousands of points across the globe. These models are run on supercomputer systems that cost tens of millions of dollars each.

The Intergovernmental Panel on Climate Change uses outputs from about 20 computer models to conclude that catastrophic global warming is on the way. On top of the 0.7°C global temperature rise over the last 130 years, the IPCC projects an additional rise of 1.5°C to 4.5°C, or an average of 3°C (5.4°F) by the year 2100. However, as we'll discuss in Chapter 5, all major models are constructed on key assumptions about climate change that do not match empirical data.

THE DOGMA OF MAN-MADE GLOBAL WARMING

Based on this simplistic science, Climatism bombards mankind with the message that "man-made greenhouse gases are destroying Earth's climate." Many terrible disasters are projected, including melting of Earth's icecaps resulting in sea level rise and immersion of coastal cities, stronger hurricanes and tropical storms, droughts and floods, species extinction, deaths from heat waves and disease, and other calamities.

A Few of Many Events Blamed on Global Warming[6]

"The Athabasca Glacier drains the Columbia ice field in the Canadian Rockies and has retreated more than a kilometer since the early twentieth century…"

"…the small icecap atop Mount Kilimanjaro in Tanzania has lost 80 percent of its mass…"

"Extensive coral communities in the Indian Ocean have been damaged permanently by warm sea temperatures, with up to 90 percent of the corals killed."

"Dramatic losses of coastal wetlands have already occurred along the Gulf of Mexico coast."

"During the 1970s and 1980s, the Sahel had a long string of drought years…"

"Rain has been falling in intense bursts in Europe more frequently in recent decades."

"The devastating European heat wave of 2003 was a truly extreme weather event by the standards of the twentieth century, and the death toll has been estimated at over 35,000 people."

"Hurricane Katrina…ranks as the most expensive natural disaster in the history of the US."

Climatism has become a dogma, a system of belief that attributes any natural change to changes in a trace gas in our atmosphere. Scientists from all over the world report on changes to icecaps, glaciers, weather, coral reefs, plant life, animal habitats, and land and sea environments, blaming all warming events on man-made causes. Advocates of human-caused warming are compelled to provide a man-made explanation for every weather event. Didn't the UK Meteorological Office just install a £30 million computer to study climate change? Then they must have an answer for the cold winter of 2011 and every other weather occurrence.

Climate science has abandoned com-

Big Green for Climate Change

Construction Begins on $100 Million Climate Supercomputer in Wyoming
"About $20 million will come from the Wyoming state government. Other funding is coming through the National Science Foundation."
—*USA Today*, June 15, 2010[7]

WACKY CLIMATE CHANGE HEADLINES

"Scientists: Aliens May Punish Our Species for Climate Change"
—*The Atlantic*, August, 2011[8]

mon sense. Natural climate change has disappeared, replaced only by man-made climate change. Dr. S. Fred Singer, Professor Emeritus of Environmental Sciences of the University of Virginia, offers sound advice to the contrary:

> Evidence of warming is not evidence that the cause is anthropogenic [man-made].[9]

The point is that warming events, which climate scientists endlessly announce, *are not evidence* that greenhouse gases, let alone man-made emissions, are the *cause* of global warming. In fact, *there is no direct empirical evidence that man-made greenhouse gas emissions are the primary cause, or even a significant cause, of global warming.*

AN OPEN AND SHUT CASE?

From these four basic premises—rising global surface temperatures, rising atmospheric carbon dioxide concentration, the greenhouse effect, and computer model projections—Climatism demands global changes in all aspects of our lives. But all is not as it seems. As we'll see in Chapters 4 and 5, Climatism has drawn several scientific conclusions that are false, and because they are false, *all four of these premises are undermined*. Let's take a deeper look at climate science.

CHAPTER 4

HISTORY SHOWS
GLOBAL WARMING *NOT* ABNORMAL

"Climate has been changing for billions of years."
—DR. BUZZ ALDRIN, NASA ASTRONAUT

Based upon the simple science that "man-made emissions cause global warming," five conclusions drive today's global climatology. But an objective analysis of empirical data shows that these conclusions are false. Nevertheless, these have become dogma, to be defended in the face of any and all empirical evidence. The conclusions are: 1) that the Modern Warming, the rise in global temperatures over the last 130 years, is abnormal, 2) that carbon dioxide drives Earth's climate, 3) that the rise in atmospheric CO_2 is primarily due to man, 4) that water vapor boosts global warming from CO_2, and 5) that the sun is an insignificant factor in global warming. We'll discuss the first conclusion in this chapter and then the others in Chapter 5.

CONCLUSION #1: MODERN WARMING ABNORMAL?

The first conclusion of Climatism is that the Modern Warming is abnormal. Rajendra Pachauri, the Chairman of the IPCC, says "the warming of the climate system is unequivocal."[1] In its Fourth Assessment Report (AR4) of 2007, the IPCC states:

> …it is *extremely unlikely* that the global climate change of the past 50 years can be explained without external forcing and *very likely* that it is not due to known natural causes alone.[2]

Table 9.4 of Chapter 9 of AR4 summarizes the IPCC evidence. The table concentrates on a host of twentieth century temperature data, greenhouse gas increases, and model results. But common sense and sound scientific reasoning demand that, if you want to know if the modern rise is abnormal, then you need to compare it to *past* temperatures. Chapter 6 of AR4 attempts to do this, but as we will show, the IPCC takes only a selective look at historical data and ignores many studies that reach other conclusions.[3]

The conclusion that the Modern Warming is abnormal is the cornerstone of global warming alarmism. If the twentieth century rise in surface temperatures is not abnormal, then what's all the fuss about?

MEDIEVAL WARM PERIOD AND LITTLE ICE AGE

A favorite phrase in the media is: "Climate change is real!" A true statement, but it's about as useful as "grass is green." Of course climate change is real. Global warming is real. Global cooling is real. *Climate change is not only real, but continuous.* It's been happening throughout all of Earth's history. Two well-known historical periods of contrasting climate are the Medieval Warm Period (MWP) and the Little Ice Age (LIA).

The MWP was a period from about AD 900–1300 when temperatures were at least as warm as those of today. The Viking Eric the Red led 25 ships and 500 settlers to found

CLOWNY'S LIST OF POWERFUL STATEMENTS

Grass is Green!
Water is Wet!
Climate Change is Real![4]

the town of Hvalsey in southwest Greenland in AD 985.[5] Over the next 300 years, the town prospered and grew to a population of 5,000 inhabitants. Hvalsey residents farmed, raised livestock, and traded polar bear skins and walrus tusks. According to the *Book of Icelanders*, southwest Greenland of that time contained birch thickets with trees six meters high due to the warm climate—quite different from the low-scrub vegetation of today.[6]

The Medieval Warm Period has been called an era of "climate optimum." Throughout history, mankind has prospered during warmer times, such as the MWP. Many of the cathedrals of Europe were started during this age, such as the great cathedral in Cologne, begun in 1248.[7] Back then, England was a producer of wines, with fourteen different vines grown

Viking Church in Hvalsey, Greenland, built in the 12th century. (Photo by Munksgaard)[8]

commercially.[9] Despite the recent warming of the twentieth century, temperatures still don't favor wine production in England.

About AD 1300, Earth's climate entered the cooler period of the Little Ice Age, which lasted to about AD 1850. Although not a true ice age, temperatures during the LIA were about 1–2°C cooler than the MWP and those of today. The settlement at Hvalsey, Greenland, was beset by shorter growing seasons and sea ice that blocked ship-borne trade. It was abandoned in the 1400s.

Colder periods have always been times of human hardship, and the Little Ice Age was no exception. European historical records describe shorter growing seasons, increased famine, and disease during the LIA. In 1695, Iceland was completely surrounded by ice, which extended southeast as far as the Faeroe Islands. The population of Iceland dropped by 50 percent during the LIA.[10]

During the Little Ice Age, the Thames River froze solid at London every year and became the site of the annual "Frost Fair." Londoners built sheds, erected tents, and drove horses and wagons onto the ice. Numerous paintings by artists of the day have captured the scene of the Frost Fair. Note that a similar event on Thames River ice is not possible today. The Thames has not frozen solid at London for over a century.[11]

The Medieval Warm Period and the Little Ice Age are examples of warm and cold climatic periods that were entirely due to natural cycles.

London Frost Fair, 1683. Fair on frozen Thames River. (Image of painting by Thomas Wyke)[12]

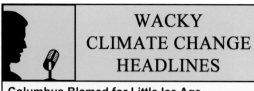

WACKY CLIMATE CHANGE HEADLINES

Columbus Blamed for Little Ice Age
"By sailing to the New World, Christopher Columbus and other explorers who followed may have set off a chain reaction of events that cooled Europe's climate for centuries."
—*ScienceNews*, October 13, 2011[13]

They were also relatively recent eras in terms of climate, which changes over hundreds or thousands of years. Yet, the IPCC discounts these natural events, claiming that they only happened in Europe and were not global, or that the temperature changes were small compared to the twentieth century warming. Let's look at scientific evidence from around the world and see if you agree.

EVIDENCE FOR THE MWP AND LIA AS GLOBAL EVENTS

Modern temperature records based on thermometer measurements stretch back only about 130 years. So how do we estimate past temperatures? Scientists use a combination of written historical records and temperature proxies. Temperature proxies are created by measurement of physical or chemical processes that change in parallel with historical temperatures. These include the width of ancient tree rings and the atomic composition of oxygen atoms (isotope ratio) in glacial ice and deep-sea sediment cores.

Dr. Håken Grudd of the University of Stockholm completed a temperature history for the Torneträsk area of northern Sweden in 2008 (top graph opposite). Using tree-ring width and density of the Scots pine as a temperature proxy, Dr. Grudd found temperatures during the Medieval Warm Period to be warmer than those of today:

> The 200-year warm period centered on AD 1000 was significantly warmer than the late-twentieth century…The new tree-ring evidence from Torneträsk suggests that this "Medieval Warm Period" in northern Fennoscandia was much warmer than previously recognized.[14]

Dr. Karin Holmgren also of the University of Stockholm, along with other researchers, constructed a temperature history from a cave in the Makapansgat Valley of South Africa. Holmgren used variations in the atomic composition of oxygen and carbon (isotopes) found in cave stalagmites to estimate past temperatures. The research paper, published in 2005, showed that peak temperatures of the MWP were more than 2°C higher than those of today.[15]

We find similar evidence in the Pacific Ocean. Dr. Delia Oppo of the Woods Hole Oceanographic Institution led a study of sea surface temperatures (SSTs) in the Makassar Strait, between Borneo and Sulawesi Indonesia, in 2009. Changes in oxygen isotopes in the

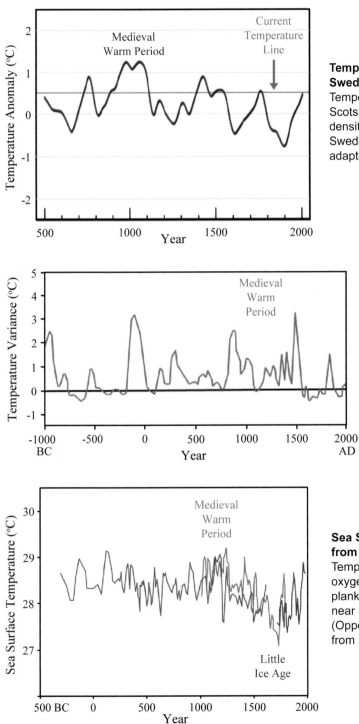

Temperature History from Sweden Tree Rings
Temperatures constructed from Scots pine tree ring width and density found in Torneträsk, Sweden. (Grudd et al., 2008, adapted from *CO₂Science*)[16]

Temperature History from Cave Stalagmites in South Africa
Temperatures constructed from oxygen and carbon isotopes of cave stalagmites in the Makapansgat Valley of South Africa. (Holmgren et al., 2001, adapted from *CO₂Science*)[17]

Sea Surface Temperature from Plankton Near Indonesia
Temperatures constructed from oxygen isotopes of shells of plankton in the Makassar Strait near Sulawesi, Indonesia. (Oppo et al., 2009, adapted from *CO₂Science*)[18]

shells of plankton that accumulated on the ocean floor were used as a proxy. The study found evidence for warm temperatures during the MWP and cooler temperatures during the LIA:

> …the reconstruction suggests that at least during the Medieval Warm Period, and possibly the preceding 1,000 years, Indonesian SSTs were similar to modern SSTs.[19]

It's been said that while Americans "think in years," Chinese "think in centuries." The Chinese people are known for their excellent historical records. Dr. Quansheng Ge of the Chinese Academy of Sciences led other researchers who assembled a 2000-year temperature record for Eastern China based on historical data, including dates for frosts and snows, the thawing of rivers and lakes, distribution and harvest times for agricultural crops, and blossoming of flowers (top graph opposite). The 2003 paper shows strong evidence for the MWP and LIA in Eastern China. Dr. Ge states:

> Starting from AD 510s, the temperature rose rapidly and entered the warm period of the AD 570s–1310s. In this epoch, climate was dominantly warmer…The 30-year mean temperatures of two warm peaks were generally 0.3–0.6°C higher than present day…[20]

Dr. Andrei Andreev of the Alfred Wegener Institute for Polar and Marine Research and others completed a study in 2005 providing a temperature record for Central Asia. Sediments were retrieved from the bottom of Lake Teletskoye in the northern Altai Mountains, with pollen and charcoal records used as temperature proxies. The constructed record shows clear evidence for a MWP that was warmer than today as well as the colder era of the LIA. Dr. Andreev reports:

> Around 1200 AD, climate became more humid with temperatures probably higher than today. This period of rather stable climate possibly reflecting the Medieval Warm Epoch lasted until AD 1410…A subsequent period with colder and more arid climate conditions between AD 1560 and 1820 is well correlated with the Little Ice Age.[21]

Julie Richey, PhD candidate at the University of South Florida, and other researchers published a temperature reconstruction for the Pigmy Basin of the northern Gulf of Mexico in 2007. Sediment cores were retrieved from the continental shelf and oxygen isotope proxies from the shells for plankton were used to estimate sea surface temperatures (SSTs) over the last 1400 years. The study found:

> …two multidecadal intervals between 1000 and 1400 yr B.P. [before present] suggest that SST was as warm or warmer than near-modern SST at that time…data also suggest SST

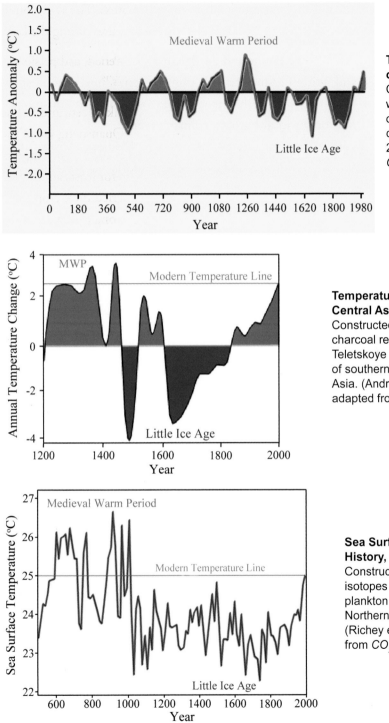

Temperature History of Eastern China
Constructed from written history records of agriculture and climate. (Ge et al., 2003, adapted from *CO2Science*)[22]

Temperature History of Central Asia
Constructed from pollen and charcoal records from Lake Teletskoye in the Altai Mountains of southern Russia, in central Asia. (Andreev et al., 2007, adapted from *CO2Science*)[23]

Sea Surface Temperature History, Gulf of Mexico
Constructed from oxygen isotopes in the shells of plankton from the Pigmy Basin, Northern Gulf of Mexico. (Richey et al., 2007, adapted from *CO2Science*)[24]

cooling during intervals of the LIA that was at least 2°C below near-modern SST.[25]

We've discussed peer-reviewed studies of tree rings in Sweden, cave stalagmites in South Africa, shells of plankton from both the Pacific and the Atlantic Ocean, carbon and pollen records from Central Asia, and written history records from Eastern China that all show a Medieval Warm Period as warm or warmer than temperatures of today. So how can the IPCC claim that the MWP was only a local European climate event?

In fact, these are just six of hundreds of studies that show the Medieval Warm Period and the Little Ice Age to be natural, significant, and global events of Earth's climate history. The Center for the Study of Carbon Dioxide and Global Change, directed by Dr. Craig Idso, provides the website *CO₂Science*, that contains extensive evidence for the Medieval Warm Period and Little Ice Age. The site lists more than 300 peer-reviewed studies documenting the MWP and the LIA from all seven continents. Of these, 91 studies enable a numerical temperature comparison of the MWP and the twentieth century warming. *Three quarters of these studies show that temperatures of the Medieval Warm Period were as warm or warmer than temperatures of today.*[26]

THE 1500-YEAR CLIMATE CYCLE

The Medieval Warm Period and Little Ice Age are just two of numerous alternating warm and cool periods that are part of Earth's climatic history. Dr. Fred Singer and Dennis Avery, a senior fellow at the Hudson Institute, wrote the excellent book *Unstoppable Global Warming: Every 1500 Years*, proposing that the MWP and LIA are part of a natural cycle in Earth's temperatures that is roughly 1,500 years long. The authors observed:

> The Earth currently is experiencing a warming trend, but there is scientific evidence that human activities have little to do with it. Instead, the warming seems to be part of a 1,500-year cycle (plus or minus 500 years) of moderate temperature swings.[27]

The 1500-year cycle appears to be a periodic 1–2°C change in Earth's temperature. These cycles are shorter and more moderate than the very-long-term cycles that drive the large temperature changes of the Ice Ages. Analysis of ice cores from Greenland clearly shows several cycles since the last ice age, including warm periods of the Holocene Climate Optimum, the Roman Climate Optimum, the Medieval Warm Period, and the Modern Warm Period, along with intervening cool eras such as the Little Ice Age.

In 2010, Dr. James Hansen at NASA issued a press release that stated "January 2000 to

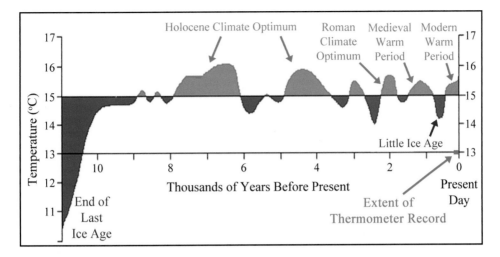

Temperature Cycles of the Last 12,000 Years. Reconstructions show temperatures in Earth's Northern Hemisphere varied by about 2°C since the end of the last ice age. Data from oxygen isotope analysis of ice cores from Crete, Greenland. Note the small extent of the modern thermometer record. (Dansgaard et al., 1984, adapted by Avery, 2009)[28]

December 2009 was the warmest decade on record."[29] This phrase was trumpeted in the news media and continues to be used by man-made global warming advocates around the world. While true, it is a very misleading statement. The "record" refers to modern thermometer measurements of global temperature, which only date back to the late 1800s, which is just a mole hill on the mountain of climate history. The statement "warmest on record" ignores all of the changes in climate for the last 12,000 years and attempts to convince the reader that recent temperatures are abnormal, in-line with Climatist dogma.

THE IPCC TEMPERATURE CURVE

The IPCC temperature curve can be described as fuzzy at best. It's been as tricky as a skateboard on wet cobblestones. The IPCC First Assessment Report of 1990 showed a 1,000-year temperature graph that

DO POLAR BEARS STILL EXIST?

"They're imaginary! They're virtual reality! They can't possibly be there! They all died out during past warmer periods."
—Geologist Bob Carter, poking fun at Climatism (Photo by Wilson)[30]

included the Medieval Warm Period and the Little Ice Age (top graph opposite). But this view of history did not fit with the dogma that the modern warming was abnormal. Dr. William Happer, Professor of Physics at Princeton University, remarks:

> The existence of the Little Ice Age and the Medieval Warm Period were an embarrassment to the global-warming establishment, because they showed that the current warming is almost indistinguishable from previous warmings and coolings that had nothing to do with burning fossil fuel. The organization charged with producing scientific support for the climate change crusade, the Intergovernmental Panel on Climate Change (IPCC), finally found a solution. They rewrote the climate history of the past 1000 years with the celebrated "hockey stick" temperature record.[31]

Indeed, the 2001 Third Assessment Report claimed to have found "new evidence" that supported the case for abnormal modern warming. On the first page of the Summary for Policymakers, it states:

> New analysis of proxy data for the Northern Hemisphere indicate that the increase in temperature in the 20th century is likely to have been the largest of any century during the past 1,000 years.[32]

The new analysis chart became the infamous "Mann Hockey Stick Curve," (center graph, opposite page). The graph is adapted from a 1999 paper by Dr. Michael Mann at the University of Massachusetts, along with other authors.[33] The curve showed a 1000-year period of little-changing temperatures (the shaft of the hockey stick) capped by an upward temperature spike in the twentieth century. The Medieval Warm Period and the Little Ice Age had disappeared! Mann used tree-rings, ice cores, and other proxies to reach his startling conclusions.

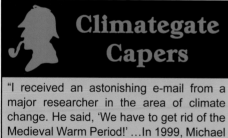

Climategate Capers

"I received an astonishing e-mail from a major researcher in the area of climate change. He said, 'We have to get rid of the Medieval Warm Period!' ...In 1999, Michael Mann and his colleagues published a reconstruction of past temperature in which the MWP simply vanished..."
—Dr. David Deming, testimony before US Senate, 2006[34]

Despite a large body of scientific evidence to the contrary, the IPCC adopted the Mann curve as proof of man-made global warming. John Houghton, then Chairman of IPCC Working Group I, made numerous presentations with a chart of the Mann Curve behind him. Al Gore and many others did the same. The Canadian government mailed a copy of the curve to every household in Canada. Governments around the world posted the Mann Curve for all to see.

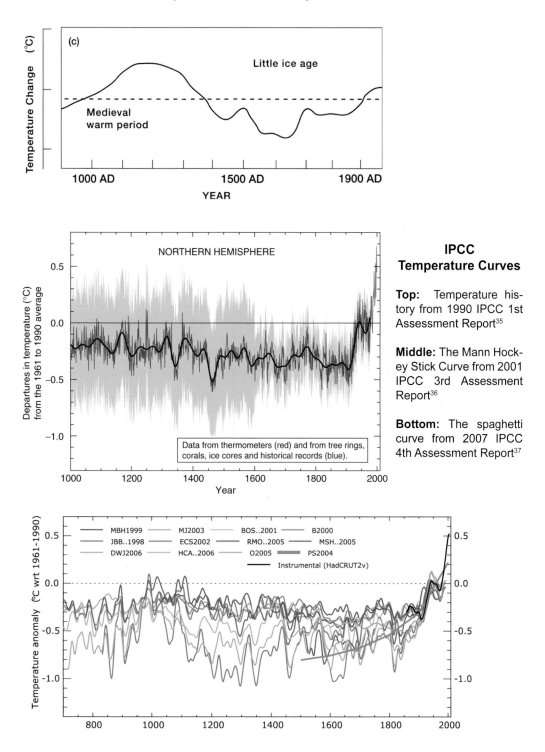

IPCC Temperature Curves

Top: Temperature history from 1990 IPCC 1st Assessment Report[35]

Middle: The Mann Hockey Stick Curve from 2001 IPCC 3rd Assessment Report[36]

Bottom: The spaghetti curve from 2007 IPCC 4th Assessment Report[37]

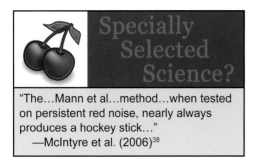

Specially
Selected
Science?

"The...Mann et al...method...when tested on persistent red noise, nearly always produces a hockey stick..."
—McIntyre et al. (2006)[38]

Two Canadians, Stephen McIntyre, author of the website *Climate Audit*, and Dr. Ross McKitrick, Professor of Economics at the University of Guelph, made repeated requests over several years to obtain Dr. Mann's data and algorithms. Reproducing the results of a study is standard practice in the scientific community. After the data was grudgingly provided, McIntyre and McKitrick found the analysis to be rife with errors.[39] In fact, *random data could be fed into Mann's algorithm and a hockey stick-shaped curve would almost always be produced!* The saga of the Mann Curve culminated in a hearing in the US House of Representatives in 2006, where Mann's work was completely discredited.

Undaunted, the IPCC produced a new temperature curve for the Fourth Assessment Report in 2007, without any explanation of what happened to the Mann Curve. It was a *spaghetti curve* of twelve proxy studies that showed a little bit of the MWP, but still claimed that the modern warming was abnormal. However, *none* of the dozens of peer-reviewed studies listed on the CO_2*Science* website that showed the MWP warmer than today was used by the IPCC for the spaghetti curve.

FAILED PREDICTIONS

"...the planet will cool, the water vapor will fall and freeze, and a new Ice Age will be born."
—*Newsweek*, January 26, 1970

"New Ice Age—It's Already Getting Colder. Some midsummer day, perhaps not too far in the future, a hard, killing frost will sweep down on the wheat fields of Saskatchewan, the Dakotas and the Russian steppes..."
—*Los Angles Times*, October 24, 1971

"A recent flurry of papers has provided further evidence for the belief that the Earth is cooling. There now seems little doubt that changes over the past few years are more than a minor statistical fluctuation."
—*Nature*, March 6, 1975[40]

SHORT-TERM CYCLES EXPLAIN MODERN WARMING

If we take a closer look at the modern warming and the rise in atmospheric carbon dioxide, we find that the curves don't track very well. Carbon dioxide increased steadily since the first measurements were made in Hawaii in 1958, but temperatures have not. Temperatures declined from about 1940 to about 1975, triggering some of the "new ice age" headlines in the 1970s. They rose from about 1975 to about 2000, providing the fodder for Climatism to create the current worldwide global warming scare. But since the beginning of our new

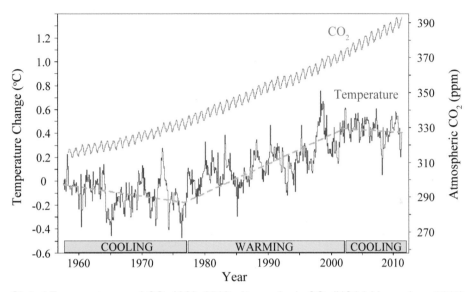

Global Temperature and CO₂ 1960–2010. Atmospheric CO_2 (NOAA Mauna Loa, 2011) and global temperatures (CRU HADCRUT3, 2011) over the last 50 years. Carbon dioxide levels have climbed steadily, while global surface temperatures have experienced cycles of warming and cooling. (Adapted from *Climate4You*)[41]

century, global temperatures have been flat to declining, despite the continuing rise in CO_2. None of the 20 climate models that the IPCC relies on predicted the declining temperature trends of the last ten years. Is something else at work here?

In fact, recent global temperature trends are dominated by short-term natural cycles. These short cycles are superimposed upon the long-term cycles that drive the ice ages, and the 1500-year cycle proposed by Singer and Avery with its Medieval Warm Period and Little Ice Age. The El Niño-La Niña fluctuation is the best known of Earth's short-term temperature cycles.

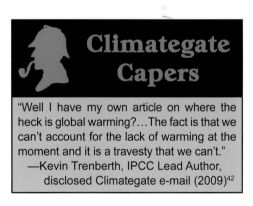

Climategate Capers

"Well I have my own article on where the heck is global warming?…The fact is that we can't account for the lack of warming at the moment and it is a travesty that we can't."
—Kevin Trenberth, IPCC Lead Author, disclosed Climategate e-mail (2009)[42]

El Niño, more completely known as the El Niño Southern Oscillation (ENSO), is a very short cycle that occurs in the southern Pacific Ocean. The lesser-known part of ENSO is named La Niña. During La Niña, trade winds blow primarily from east to west across the

Pacific Ocean. Storms form above the warm pool of water in the western Pacific. Cold water from the deep ocean wells up near Peru, bringing nutrients to Peruvian fishing waters.

Then, every few years, the El Niño part of the cycle takes over. A major temperature shift occurs in the Pacific Ocean, pushing a huge amount of heat thousands of miles to the east. Since the Pacific Ocean is the largest heat reservoir on Earth, El Niño affects weather worldwide. The Pacific trade winds change direction and blow from west to east. Less cold water wells up near Peru, reducing fishing harvests. Storms form in the central Pacific and move east to hit California, causing flooding and mud slides. Weather conditions change in Africa, India, Australia, and all over the world. El Niño even reduces hurricane activity in the Atlantic Ocean.[43]

Scientists don't know what exactly causes the ENSO cycle, but evidence shows that El Niños have been occurring for thousands of years. ENSO is just one of many short-term natural cycles of Earth that scientists are now studying, including the Pacific Decadal Oscillation, the Atlantic Multidecadal Oscillation, and cycles in the Arctic, the Antarctic, and the Indian Ocean. Each of these cycles has been operating for thousands of years, long before any man-made greenhouse gas emissions.

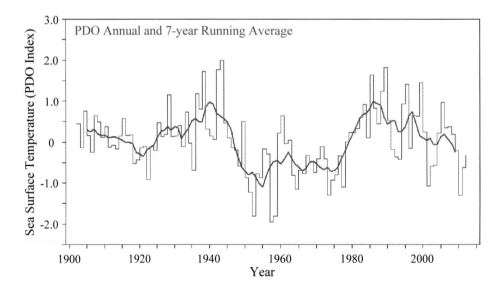

The Pacific Decadal Oscillation 1900–2010. The variation in northern Pacific Sea Surface Temperatures (the PDO Index) for annual and seven-year running average (dark line) over the last 110 years. (University of Washington, JISAO, 2011)[44]

The Pacific Decadal Oscillation (PDO) is of special interest, because it is probably the most powerful of Earth's short-term cycles. The PDO was named by fisheries scientist Dr. Stephen Hare in 1996 to describe the relationship between Alaskan salmon harvests and climate in the north Pacific Ocean. The PDO is a cycle of sea surface temperatures in the northern Pacific that shows a temperature change of about plus or minus two degrees Celsius over a period of about 50 years.[45] The very short-term changes of El Niño occur superimposed on top of the PDO cycle. Note from the graph that the change in PDO index generally tracks the variation in global temperatures over the last century, particularly the cooling from 1940–1975, the warming from 1975–2000, and the recent cooling of the twenty-first century.

Recall that the IPCC claims that global temperatures over the last 50 years cannot be explained by natural causes. Yet, even a cursory look at Earth's temperature cycles shows this assertion to be false. Dr. Syun Akasofu, Professor of Physics at the University of Alaska, disagrees with the IPCC. Akasofu proposes that recent twentieth century temperatures are easily explained by a combination of a long-term warming and short-term temperature variation. The 0.7°C rise in average temperatures since 1880 is part of a long-term warming as Earth recovers from the Little Ice Age. The rapid rise from 1975 to 2000 is due to the additive effect of a rising cycle of the Pacific Decadal Oscillation. Dr. Akasofu comments:

> …it is quite obvious that the temperature change during the last 100 years or so includes significant natural changes, both the linear change and fluctuations. It is very puzzling that the IPCC reports state that it is mostly due to the greenhouse effect.[46]

THE MODERN WARMING IS NOT ABNORMAL

So the basis for all the alarm, the temperature rise of the last years of the twentieth century, is not abnormal in terms of the size of the change, the rate of the change, or the levels of measured temperatures. Proxy and historical data show that temperatures during the Medieval Warm Period and other recent eras in Earth's climate history were warmer than those of today. In addition, the Pacific Decadal Oscillation and other temperature cycles fully explain the modern warming, without need for the claimed impacts from man-made greenhouse gases. The Climatist conclusion that the Modern Warm Period is abnormal is not supported by historical evidence. Let's look at four other major false conclusions in the next chapter.

CHAPTER 5

CLIMATE SCIENCE—THE REST OF THE STORY

*"It doesn't matter how beautiful your theory is, it doesn't matter how smart you
are. If it doesn't agree with experiment, it's wrong."*
—NOBEL PRIZE WINNING PHYSICIST RICHARD P. FEYNMAN

Climatists ask skeptics: "Don't you agree that man-made emissions contribute to
the greenhouse effect?" The answer is yes, of course. So according to Climatism,
the fate is accomplished, the case is closed, and the debate is over. If emissions
contribute to global warming, they must be stopped.

The problem with this logic is that the size of the contribution is critical. As an illustrative example, consider that house cats kill millions of birds each year, so they contribute to species extinction. Our planet would not have as many house cats if they weren't favored by humans as pets. But can we stop species extinction by banning house cats? Of course not. Just as in the case of the climate, the size of the contribution of man-made emissions is critical. Last chapter we showed that the rise in global temperatures during the twentieth century was not an abnormal climate event, as Climatists have concluded. Let's examine four additional shaky conclusions.

CONCLUSION #2: CO_2 DRIVES EARTH'S CLIMATE?

The second false conclusion of Climatism is that carbon dioxide, a trace gas in our atmosphere, drives Earth's climate. After years of repetition of global warming doctrine, much of climate science now views the world through carbon-tinted glasses. In his book *Storms of My Grandchildren*, Dr. James Hansen claims that changes in CO_2 levels caused the start and end of ice ages, the rise and fall of the seas, and even most extinctions of animal species throughout millennia. He discounts the effects of "insignificant" factors like the sun and the planets, and ignores the titanic forces of Earth's weather and oceans, stating:

> …the global surface albedo [surface whiteness] and greenhouse gas changes account for practically the entire global climate change.[1]

This is an astonishing conclusion. Climate science has jumped off the bridge over CO_2 . The small contribution of carbon dioxide to the greenhouse effect, just one of many physical processes of Earth, has become the explanation for every global event, even earthquakes. After building computer models to show that the twentieth century warming of one degree was due to CO_2 increase, Climatists now conclude that Earth's climate throughout all of history was driven by CO_2 . But in fact, carbon dioxide is only a small part of Earth's climate.

Earth's atmosphere is composed of 78 percent nitrogen gas, 21 percent oxygen, and 1 percent of other trace gases. Carbon dioxide is one of the trace gases, comprising much less than 1 percent. *Only four of every ten thousand air molecules are carbon dioxide.* In all of human history, man-made emissions are responsible for adding *only a fraction of one* of these four molecules.

Earth's climate is amazingly complex. It's driven by gravitational forces of our solar system, radiation from the sun, and cosmic rays from stars in deep space. Climate is a chaotic, interdependent system of atmosphere, biosphere, land, ocean, and deep ocean. It's been changing through cycles of warming and cooling, tropical ages, temperate ages, and ice ages throughout all of Earth's history. The "simplified" diagram opposite shows only a few of hundreds of forces involved.

Energy from the sun drives all weather on Earth. Sunlight strikes the Equator more

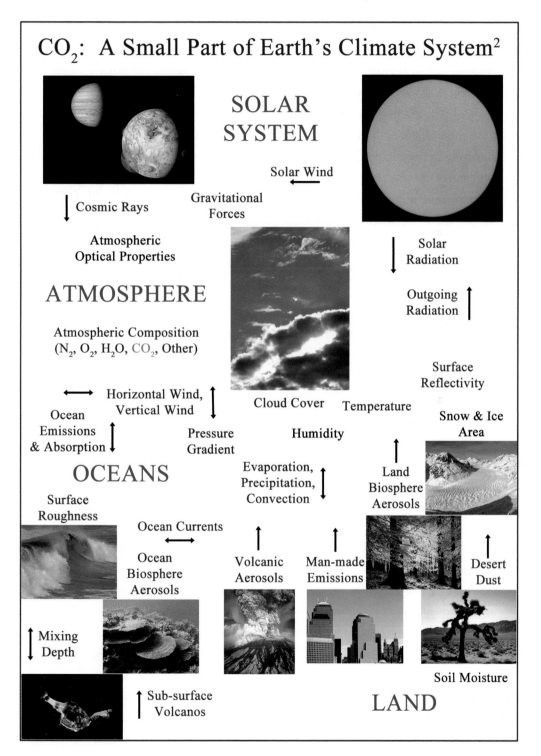

CO_2: A Small Part of Earth's Climate System[2]

SOLAR SYSTEM

Solar Wind

Cosmic Rays

Gravitational Forces

Atmospheric Optical Properties

Solar Radiation

Outgoing Radiation

ATMOSPHERE

Atmospheric Composition (N_2, O_2, H_2O, CO_2, Other)

Surface Reflectivity

Horizontal Wind, Vertical Wind

Cloud Cover   Temperature

Snow & Ice Area

Ocean Emissions & Absorption

Pressure Gradient

Humidity

OCEANS

Evaporation, Precipitation, Convection

Land Biosphere Aerosols

Surface Roughness

Ocean Currents

Ocean Biosphere Aerosols

Volcanic Aerosols

Man-made Emissions

Desert Dust

Mixing Depth

Sub-surface Volcanos

Soil Moisture

LAND

directly than it does the North and South Pole, so more energy is absorbed in the tropics. Ocean currents, storm fronts, cyclones, the jet stream, and trade winds are all part of a weather system that redistributes heat energy from the tropics to the polar regions. These powerful forces shape our climate.

The oceans have a huge impact on Earth's temperatures. The Gulf Stream ocean current in the Atlantic Ocean is the dominant driver of weather and temperature in Europe. The El Niño cycle in the Pacific Ocean affects weather all over the world. Our oceans contain over 250 times the mass of the atmosphere and can hold about 1,000 times more heat.[3] But Climatists are obsessed with the greenhouse effect, the warming from the absorption of outgoing infrared radiation by the trace gas carbon dioxide.

Additional factors, such as gases from volcanic eruptions, pollen and other aerosols emitted from vegetation and plankton, and desert dust, also shape our climate. Renowned physicist Freeman Dyson of Princeton University comments on the complexity of the climate:

> …the computer models are very good at solving equations of fluid dynamics but very bad at describing the real world. The real world is full of things like clouds and vegetation and soil and dust which the models describe very poorly.[4]

It's remarkable that Climatism focuses on carbon dioxide as the cause of the twentieth century warming, and discounts the solar, weather, and ocean forces that are the dominant drivers of climate change.

CONCLUSION #3: CO_2 RISE DUE TO MAN?

The third false conclusion of Climatism is that the recent rise in atmospheric carbon dioxide is primarily due to man. Proponents of man-made global warming agree that yes, carbon dioxide is a trace gas, but then they argue that the rise in atmospheric concentration has been unprecedented.

The IPCC uses a graph showing a spike in recent CO_2 levels from flat lower levels in past thousands of years. The purpose of this graph is two-fold. First, it's a highly compressed version of the Keeling Curve on page 49, intended to show that the rise in CO_2 is alarming. Second, the graph proposes that the rise is abnormal in history, and therefore *primarily due* to man-made emissions. It's interesting to see that, on a different graph where one percent of the atmosphere is used on the vertical axis instead of a truncated scale in parts per million, the recent rise in CO_2 can hardly be seen.

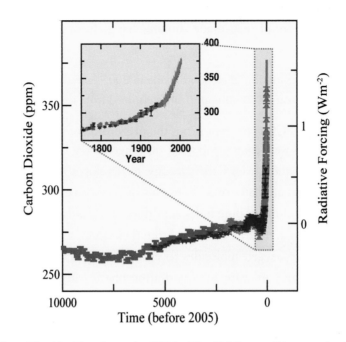

Carbon Dioxide Rise from the IPCC. The IPCC uses this curve to claim the rise in CO_2 to be unprecedented in the last 10,000 years. Measurements shown are from ice cores (symbols with different colors from different studies) and atmospheric samples (red lines). (IPCC AR4, 2007)[5]

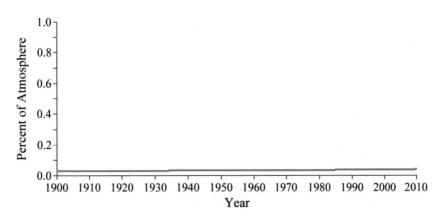

The "Terrible Rise" in Carbon Dioxide 1900–2010. A portion of the same rise in atmospheric CO_2 shown in the graph above, but using a scale with one percent of the atmosphere as the highest value. Carbon dioxide is a trace gas.

The claim that atmospheric carbon dioxide was unchanging for thousands of years is fundamental to the theory of man-made climate change from greenhouse gas emissions. The IPCC states:

> The atmospheric concentration of carbon dioxide in 2005 exceeds by far the natural range over the last 650,000 years (180-300 ppm) as determined from ice cores.[6]

The problem with the IPCC statement is not the magnitude of the recent rise in CO_2, but the claim that CO_2 levels were confined to a narrow range for the last 650,000 years.

Scientists use information from Earth's icecaps to develop a record of our geologic history. Annual snowfall accumulates over thousands of years, changes to hard ice pack, and traps a record of Earth's climate. Bore holes are drilled thousands of feet into the icecaps of Antarctica and Greenland to recover ice cores. From these cores, scientists look at the isotopes (atomic composition) of oxygen molecules to estimate temperature changes. Trapped air bubbles in the ice also give an estimate of past atmospheric carbon dioxide concentration. But the ice core air bubble data, which show an unchanging "background level" of about 280 ppm prior to the twentieth century increase, are plagued with uncertainty.

Measuring the composition of air in trapped bubbles in ice is a complex process. Holes are drilled deep into the ice and cylindrical cores are removed from the bore hole. The ice cylinders can be directly analyzed to measure the ratio of oxygen isotopes, a proxy for temperature, and to estimate the age of the ice sample. But the air bubbles trapped in the ice do not have the same age as the surrounding ice. This is because air bubbles are only

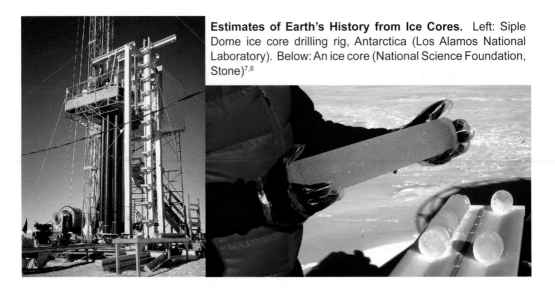

Estimates of Earth's History from Ice Cores. Left: Siple Dome ice core drilling rig, Antarctica (Los Alamos National Laboratory). Below: An ice core (National Science Foundation, Stone)[7,8]

trapped 100-200 years after snow deposition at an ice depth of about 75 meters. Above this depth, air is able to diffuse between the atmosphere and the solidifying ice. Computer models are used to estimate the age of trapped air bubbles, which is up to 200 years later than the age of the ice holding the bubbles. So, to estimate the CO_2 concentration in the atmosphere in 1850–1950, scientists analyze bubbles in ice first deposited in snowfall at about the year 1750, a process involving a lot of guesswork.[9]

In addition to the problem of determining the age of trapped air, many chemical processes active within ice can change the composition of the air bubbles. Liquid water under pressure is present in ice down to a temperature of -73°C, well below freezing. Differences in water solubility between CO_2 and other trapped gases, reactions with impurities, and the formation of clathrates (gas and ice crystals), can alter the concentration of CO_2 in the air bubbles. These processes continue to change the CO_2 concentration as long as the air bubbles are trapped in the ice.[10]

Drilling and removal of the ice cores also impacts the accuracy of the measured CO_2 concentration. Drilling causes micro-cracks to form in the cores. Drilling fluid, used to prevent freezing in the bore hole, introduces lead, zinc, sodium, and aluminum contaminants into the core samples. These elements can react with the CO_2 in the air bubbles. The micro-cracks and contaminants act to alter the CO_2 concentration in measured bubbles.[11]

Despite these measurement problems, the IPCC and the proponents of man-made warming have affirmed the ice core results and concluded that recent levels of atmospheric CO_2 are higher than at any time for hundreds of thousands of years. But other scientific studies show that this conclusion may be incorrect. The stomata of leaves tell a different story.

Stomata are small holes in leaves or pine needles of plants. They allow the plant to absorb carbon dioxide from the air for photosynthesis. Water vapor and oxygen exit the leaf through each stoma as part of the process. Scientists know that the number of stomata in leaves varies as atmospheric CO_2 changes. When the CO_2 level is high, plants grow with fewer stomata on each leaf. This allows the plant to reduce the loss of water vapor, thereby increasing drought resistance, while still able to absorb needed carbon dioxide. When CO_2 is low, leaves grow with more stomata. By counting the stomata on leaves recovered from bore holes at land drill sites, atmospheric CO_2 can be estimated for past centuries.

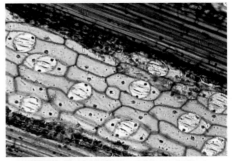

Leaf stomata of the spiderwort plant. Magnification 100 times. (Byres, FSCJ)[12]

In 2004, Dr. Lenny Kouwenberg of the University of Utrecht published data based on stomata counts showing significantly different recent levels of atmospheric CO_2 than the ice core results. CO_2 was estimated to be higher, up to 400 ppm, compared to the maximum ice core estimate of 300 ppm. In addition, changes in the levels of atmospheric CO_2 exceeded 50 ppm per century, a change that, according to climate models, could only happen due to man-made emissions.[13] A 2008 study by Dr. Thomas van Hoof and others, also of Utrecht University, showed similar rapid CO_2 changes reflected in the stomata of English Oak tree leaves. Both Kouwenberg and van Hoof concluded that the estimated carbon dioxide changes were probably caused by emissions from the oceans.[14]

Estimating atmospheric carbon dioxide concentration from leaf stomata requires careful analysis, since stomata density can be affected by factors other than CO_2, such as water availability, light intensity, and altitude. But stomata analysis offers several advantages over ice core data. Once leaves are formed and the tree dies, the stomata density is permanently fixed and unaffected by the burial process. In comparison, air in ice core bubbles is subject to chemical changes over time, as well as during recovery by a drilling team.

Stomatal analysis can also be calibrated using today's tree leaves and current atmospheric

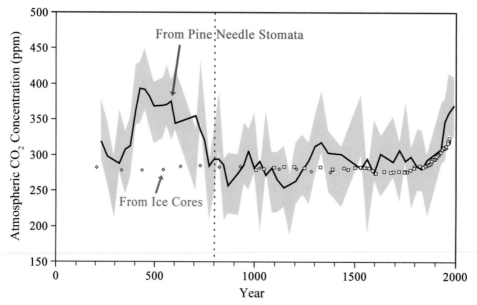

Atmospheric CO_2 from Pine Needle Stomata (300–2000). Estimates of CO_2 from the stomata of conifer pine needles retrieved from peat and lake deposits in Washington, USA (solid curve), compared to estimates from Law Dome ice core analysis. (Kouwenberg, 2004)[15]

CO_2 levels. Ice core bubbles cannot be easily calibrated to current conditions, because air bubbles are not trapped for 100-200 years until the layers of ice have hardened. It appears that air bubbles in the ice may actually measure the CO_2 concentration over many decades, providing a "smoothing process" that does not record short-term fluctuations in the level of atmospheric CO_2. In addition, it may be that CO_2 levels in trapped ice bubbles decline over time, so that ice cores understate the level of CO_2.

The IPCC Fourth Assessment Report published in 2007 fails to mention the Kouwenberg study and other stomata research used to estimate CO_2 levels. The findings of these studies conclude that atmospheric CO_2 varies due to natural emissions from the oceans, in a manner unexplained by the computer models. These research papers provide powerful evidence that the conclusion that the recent CO_2 rise is primarily due to man is false.

If we look back further in history, geologists tell us that Earth's climate over the last 400,000 years was dominated by four ice ages. Each ice age lasted about 90,000 years, separated by warm periods of about 15,000 years in length. We are currently enjoying one of these warm periods, which began about 12,000 years ago. During the last ice age, much of Northern Hemisphere was covered with an ice sheet several thousand feet thick, including the locations of New York City, Chicago, and London.

Data from the ice cores of the Vostok ice station in Antarctica show the four ice ages and intervening warm periods. Ice age transitions are associated with major changes in atmospheric CO_2 levels, which appear even with any smoothing effect caused in the trapping of air bubbles. The data also shows a strong correlation between temperature and

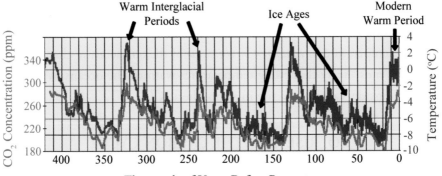

CO_2 and Temperature for the last 400,000 Years. Temperature (blue), CO_2 concentration (red), and ice ages and interglacial periods shown. Data from Vostok ice cores from Antarctica. (Adapted from Petit et. al., 1999)[16]

Shocking Discovery!

Scientists Find that Much of the Northern Hemisphere was Covered by Ice 20,000 Years Ago!

"According to researchers, global cooling and warming cycles of 7–12°C appear in the geologic record during the ice ages. They are now seeking archaeological evidence for SUVs and coal-fired power plants that could have caused these changes."[17]

CO_2 concentration. Al Gore used this correlation in his book and movie to suggest that carbon dioxide was the cause of past temperature changes.

Dr. James Hansen co-authored a 1990 paper, concluding that:

> ...the contribution of greenhouse gases to the Vostok temperature changes can be... between a lower estimate of 40% and a higher estimate of 65%.[18]

In other words, Dr. Hansen claimed that the ice ages were largely caused by changes in atmospheric carbon dioxide concentration. Up until the rise of Climatism, the cause of the ice ages was believed to be due to the Milankovich Cycles, a theory developed by the Serbian astronomer Milutin Milankovich in the 1920s. Milankovich proposed that ice ages were caused by variations in the tilt and precession of Earth's axis and the changes in the shape of Earth's orbit, driven by solar system gravitational forces.[19]

But a closer look at the ice core data shows that increases in carbon dioxide *lag* the temperature increase. Dr. Hubertus Fischer of the University of California demonstrated in 1999 that:

> High resolution records from Antarctic ice cores show that carbon dioxide concentrations increased...600±400 years *after* the warming...[emphasis added].[20]

Mr. Gore and Dr. Hansen have mixed up cause and effect. Since carbon dioxide rose centuries *after* temperature, *it could not be the cause* of the temperature change resulting in the ice ages. It's most likely that rising temperatures warmed the oceans, which then outgassed carbon dioxide, causing the rise in atmospheric CO_2.

CO_2 RISE AND THE CARBON CYCLE

In fact, most atmospheric carbon dioxide is from natural sources. Today, *50 times more* CO_2 is dissolved in Earth's oceans than is found in the atmosphere. The oceans continuously exchange CO_2 with the atmosphere, both emission and absorption, just as CO_2 escapes to the air from an opened carbonated beverage. When plants grow, they absorb

CO_2, and when they die, CO_2 is released. Volcanoes belch large quantities of CO_2 and other gases into the environment, both above and below the surface of the ocean. As the IPCC's own Carbon Cycle Model shows, only about three percent of the carbon dioxide that enters the atmosphere each year is due to human emissions.

However, the IPCC claims that carbon dioxide is "accumulating in the atmosphere" and that CO_2 stays in the atmosphere a long time:

> About 50% of a CO_2 increase will be removed from the atmosphere within 30 years, and a further 30% will be removed within a few centuries. The remaining 20% may stay in the atmosphere for many thousands of years.[21]

But this alarming claim doesn't pass the common-sense test. First, the Carbon Cycle Model estimates that, of the 750 billion tons of carbon in the atmosphere, over 200 billion tons, or about 30 percent, is absorbed by the ocean and land biosphere *each year*.[22] Second, prior

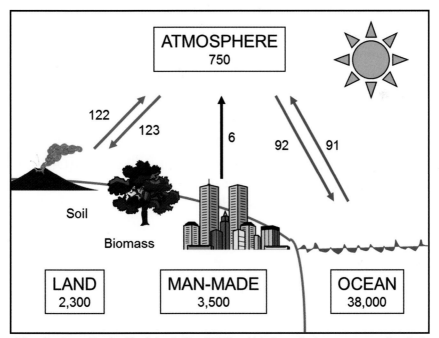

The Carbon Cycle Model of the IPCC. Numbers in boxes are estimated carbon totals residing in each climate subsystem. The numbers next to arrows are estimates of annual transfers of carbon in the form of CO_2. All numbers in billions of tons of carbon. Multiply by 44/12 to find billions of tons of CO_2. Man-made emissions are only about 3% of the CO_2 exchanged with the atmosphere each year. (Adapted from IPCC, 2007)[23]

to the advent of IPCC dogma, some 30 peer-reviewed scientific papers estimated CO_2 atmospheric lifetimes to be an average of about five to six years.[24]

Third, it's important to know that the proportion of the CO_2 in the atmosphere and oceans is governed by Henry's Law. Henry's Law is a tried-and-true law about the behavior of gases that was developed by English chemist William Henry in 1803. It states:

> For a given temperature, the amount of gas dissolved in a solution is directly proportional to the pressure of the gas in the air above the solution.[25]

This means that the CO_2 in the atmosphere and oceans should be in rough equilibrium. If CO_2 is added to the atmosphere, more will be absorbed by the oceans. Today's ocean-to-atmosphere equilibrium is roughly 50 parts of CO_2 dissolved in the ocean for each part of CO_2 in the atmosphere. Therefore, if Henry's Law holds, it's probably not possible for mankind to double the level of atmospheric CO_2, because ocean CO_2 would then also need to double to maintain the equilibrium.

Note that Henry's Law states "for a given temperature." As the temperature of the system increases, additional CO_2 molecules will gain enough energy to escape from the ocean to the atmosphere. Peer-reviewed papers estimate that CO_2 outgassing from warming oceans caused the rise in CO_2 after each ice age. The Kouwenberg thesis proposed that ocean outgassing caused CO_2 rises seen in the stomata data over the last 1500 years. As we discussed in Chapter 4, Earth has been warming for at least the last 200 years as temperatures recovered from a natural climatic cooling called the Little Ice Age. So, how much of the atmospheric rise in CO_2 does the IPCC say is due to ocean outgassing? None! The IPCC claims that the rise is due to human emissions.

If Henry's Law is in operation, man-made emissions of CO_2 would displace CO_2 outgassing from the oceans, maintaining the equilibrium of the system. The annual six billion tons of man-made emissions in the Carbon Cycle Model must be viewed relative to the more than 40,000 billion tons of carbon in the climate system, not the much smaller 750 billion tons in the atmosphere. Therefore, human emissions are only a very small part of the carbon cycle, and the IPCC alarm is misplaced.

It's important to note that the IPCC Carbon Cycle Model is fraught with large uncertainties and many assumptions. Of the flow arrows in the Carbon Cycle diagram, only the man-made 6-gigaton arrow is known with any accuracy. The 90-gigaton exchange between the oceans and atmosphere and the 122-gigaton exchange between the biosphere and atmosphere are only computer model estimates, supported by measured data made at sporadic locations across the globe, mixed with a healthy helping of Climatist dogma.

In summary, the assumption that the atmospheric rise in CO_2 is man-made does not account for important scientific data and physical laws. Leaf stomata data show large atmospheric CO_2 variations

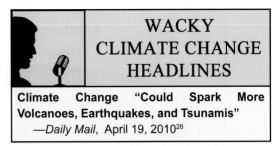

WACKY CLIMATE CHANGE HEADLINES

Climate Change "Could Spark More Volcanoes, Earthquakes, and Tsunamis"
—*Daily Mail*, April 19, 2010[26]

prior to man's industrial age. Ice core data show ice age temperature changes centuries before atmospheric CO_2 changes, indicating ocean outgassing as the possible source. Henry's Law dictates that carbon dioxide in the oceans and atmosphere be in rough equilibrium, meaning that man-made emissions are only a small part of the total carbon cycle.

SOME PERSPECTIVE ON THE GREENHOUSE EFFECT

Climatists sometimes ask, "Don't you believe in the simple physics of the greenhouse effect?" Dr. David Wojick, specialist in mathematical logic and conceptual analysis, provides a realistic answer:

> Simple physics says that if I drop a ball and a feather they will fall at the same rate. In reality, my feather blew up into a tree. It is not that the simple law is false, just that there are a number of other simple laws opposing it. In the case of climate we don't even know what some of these other laws are, so we can't explain what we see.[27]

Let's pause from our discussion of assumptions to look at a more complete discussion on the greenhouse effect and weather.

What's nature's most abundant greenhouse gas? In Chapter 2, we quoted the Sierra Club website, which called carbon dioxide the "most common greenhouse gas," which is not correct. The answer is *water vapor!* Air typically contains 500 times more water vapor than CO_2. Our previous breakdown of the percentage of gases in the atmosphere did not include water vapor. Depending upon local humidity, water vapor can form about two percent of the atmosphere, serving as the Earth's dominant greenhouse gas. Most scientists estimate that water vapor causes between 75 percent and 90 percent of greenhouse warming. If we use the conservative number, about 75 percent of the greenhouse effect is due to water vapor and clouds, and of the remaining portion, about 19 percent is due to carbon dioxide, with 6 percent due to methane and other gases.

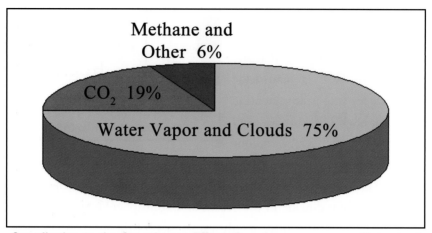

Contribution to the Greenhouse Effect. Greenhouse gases constitute about 1–2 percent of Earth's atmosphere. Of these, water vapor and clouds cause at least 75 percent of greenhouse warming. (Schmidt et al., 2010)[28]

But, as we discussed earlier, only about three percent of the CO_2 placed into the atmosphere each year is from human emissions. So, of the last 25 percent of the greenhouse effect that is due to CO_2 and methane, only about 3 percent of this is due to man-made sources. A little arithmetic shows that **mankind is responsible for only about one percent of the greenhouse effect!** When did you ever hear this figure in the news media? All we hear is how man-made emissions are causing the greenhouse effect. Yes, but human emissions are causing only about one part in one hundred of it.

In addition, radiation and the greenhouse effect are only a portion of the processes that warm or cool Earth's surface. We've added a cloud to our greenhouse effect diagram (opposite) to represent the powerful forces of weather. Weather cools Earth's surface, or if weather is less active, allows Earth's surface to warm. Evaporation is a major cooling process of our planet. Water absorbs heat from the surface as it evaporates. It then rises in the form of water vapor and condenses into clouds, releasing heat into the upper atmosphere as it condenses. Air is warmed at Earth's surface. Convection then moves heat from the surface as warm air rises, to be replaced by sinking cool air from the atmosphere. As heat reaches the upper atmosphere, radiation takes over to remove heat from the atmosphere into space. In the tropics, evaporation and convection are the primary processes removing heat from Earth's surface, and radiation is secondary.

Clouds and weather are closely connected to the greenhouse effect. When water, a greenhouse gas, evaporates, the local greenhouse effect increases. When it rains, water

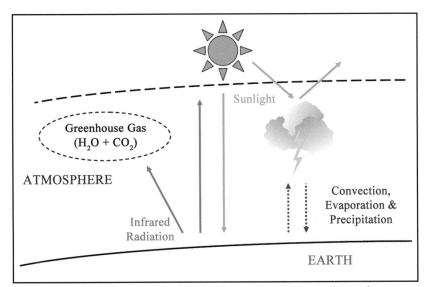

Warming and Cooling of Earth's Surface. The weather effects of convection, evaporation, and precipitation are major forces involved in warming and cooling Earth's surface. Radiation and the greenhouse effect are only part of the process.

vapor is removed from the atmosphere, and the greenhouse effect decreases. A one percent increase in global rainfall efficiency would compensate for any recent warming from CO_2. In addition, clouds reflect sunlight. About 30 percent of the sunlight that enters Earth's atmosphere is reflected by clouds. A small decrease in low-level cloudiness could be the reason for the late twentieth century warming.

Dr. Roy Spencer, principal research scientist at the University of Alabama in Huntsville, remarks:

Al Gore likes to say that mankind puts 70 million tons of carbon dioxide into the atmosphere every day. What he probably doesn't know is that mother nature puts 24,000 times that amount of our main greenhouse gas—water vapor—into the atmosphere every day and removes about the same amount every day. While this does not "prove" that global warming is not manmade, it shows that weather systems

Shocking Discovery!

Water Vapor is Nature's Most Abundant Greenhouse Gas, Say Scientists

"In amazing new findings, researchers learned that water vapor, not CO_2, is the most prevalent greenhouse gas in Earth's atmosphere. Since water vapor exhaust from industry contributes to the greenhouse effect, the Environmental Protection Agency is considering whether to declare water a pollutant under the Clean Air Act."[29]

have by far the greatest control over the Earth's greenhouse effect, which is dominated by weather and clouds.[30]

In conclusion, man's contribution to the greenhouse effect is about one percent of the total, but when we also consider evaporation, convection, and the forces of weather, human influences are an even smaller share than one percent of the total heating or cooling of Earth's surface. This means that *if we halted all man-made emissions, we might not even be able to measure the change in Earth's temperatures.*

CONCLUSION #4: POSITIVE FEEDBACK FROM WATER VAPOR?

In any reasonable scenario, carbon dioxide, by itself, can't cause catastrophic global warming. This is because absorption of infrared radiation by CO_2 is nonlinear. The first 20 parts per million of carbon dioxide in our atmosphere account for about one-half of the greenhouse warming from CO_2. Just as adding more blankets to the first blanket on your bed provides a diminishing warming benefit, adding more CO_2 to the atmosphere has a declining warming effect. Doubling of atmospheric CO_2 from 280 ppm to 560 ppm, whether from natural or man-made causes, would by itself increase Earth's surface temperature by only 1.2°C. It will take an additional doubling from 560 ppm to 1120 ppm (if not prevented by Henry's Law) to increase temperatures by another 1.2°C.[31] So how do the climate models reach their

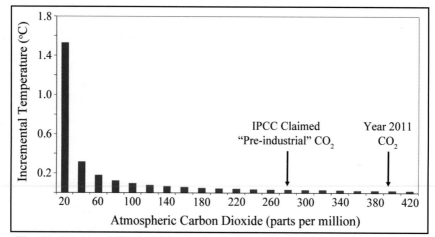

The Diminishing Effect of Atmospheric CO_2. Change in global temperature for each 20 ppm increase in atmospheric carbon dioxide. Most temperature change is caused by the first CO_2 added to the atmosphere. (Adapted from Archibald, 2008)[32]

alarming conclusions? They assume that water vapor will cause additional warming.

The fourth false conclusion of Climatism is positive feedback from water vapor. Water vapor is expected to amplify the greenhouse warming effect of carbon dioxide. The concept is that, since warmer air can hold more moisture, water vapor will increase in the atmosphere as Earth warms. Since water vapor is a greenhouse gas, a "positive feedback" from water vapor is projected to add additional warming to that caused by CO_2.

The assumption of positive feedback from water vapor has been part of climate models since the 1960s. Dr. Syukuro Manabe of the Geophysical Fluid Dynamics Laboratory in Washington, D.C., developed one of the first climate models. As part of his model, he assumed that global relative humidity remained constant as the atmosphere heated up.[33] This meant that the atmosphere would hold increasing amounts of water vapor, adding additional greenhouse heating to that of carbon dioxide.

Climate scientists use the term "climate sensitivity" to discuss how sensitive global surface temperatures are to a change in atmospheric carbon dioxide levels. The IPCC asks specifically, "If CO_2 doubles from the 280 ppm background level to 560 ppm, how much will surface temperatures rise?" The answer depends on how other systems of Earth, such as clouds and weather, respond to such a change.

All major climate models assume positive feedback, but their estimates vary. The IPCC Third Assessment Report of 2001 predicted a temperature rise of between 1.5°C and 4.5°C with a middle value of 3.0°C from a doubling of CO_2.[34] Note that for a three-degree Celsius rise in temperature, 60 percent of the model-predicted increase is due to a positive feedback greenhouse effect from water vapor, and only 40 percent is from carbon dioxide. The range in IPCC temperature estimates is primarily because each model assumes a different level of climate sensitivity and a different level of positive feedback.

So is Earth's climate sensitive to a rise in atmospheric carbon dioxide? Or to put it another way, if CO_2 rises, will the climate respond with a large positive feedback from water vapor? Or could the feedback even be negative, meaning that climatic systems act to reduce the CO_2 effect on global temperatures? An increasing body of data indicates that climate sensitivity is low.

First, empirical evidence that atmospheric water vapor is rising is sketchy, at best. Satellite data from the International Satellite Cloud Climatology Project shows atmospheric water vapor to be relatively constant over the last 30 years. Like global temperature, the trend in global water vapor is difficult to measure, but the rise in water vapor predicted by the models does not appear.

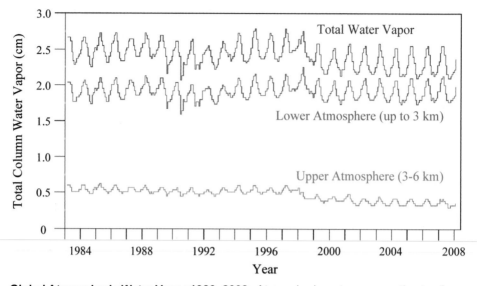

Global Atmospheric Water Vapor 1983–2008. Atmospheric water vapor estimates from satellite data. The red line is the upper atmosphere, the green line the lower atmosphere, and the blue line is the total atmosphere. No trend of increasing water vapor can be seen. The step change in 1998–1999 may be due to a change in analysis procedure. (Adapted from *Climate4You*, data from ISCCP, 2009)[35]

Second, recent peer-reviewed studies have challenged the positive-feedback water vapor assumption. That is, the effect of water vapor and clouds may act to reduce warming from rising atmospheric CO_2, rather than adding to the warming. Recent studies by Lindzen and Choi in 2011[36] and Spencer and Braswell in 2010,[37] both based on satellite data, show that climate system feedbacks are likely to be low or even negative, rather than positive. These studies estimate that an increased greenhouse effect from rising carbon dioxide is *reduced* by about one-half by reaction from the climate system. The reason for this isn't yet clear. One possible explanation is that rising temperatures create more clouds in the tropics, boosting reflection of sunlight, and causing a counter-trend cooling of Earth's surface.

Systems with positive feedback are inherently unstable. Does it make sense that Earth's climate would go off the rails from a small increase in a trace gas? In the absence of any feedbacks, doubling of atmospheric carbon dioxide will add about one percent to the sun's heating of Earth's surface. James Hansen warns of "tipping points" and "points of no return" from this tiny increase. If system feedbacks were positive, wouldn't Earth's climate have suffered runaway warming in ages past when atmospheric carbon dioxide levels were higher?

The IPCC itself has offered a way to test the models. The 2007 Assessment Report talks about a "signature" warming that should occur in the atmosphere over the equator. All major climate models predict a faster warming in the upper troposphere than at Earth's surface in the tropics. Simply put, water vapor should be rising in the upper troposphere due to the greenhouse effect, creating a temperature hot spot, shown in the top figure on the next page. But measured data from weather balloons (radiosondes) and satellites, shown in the bottom figure on the next page, does not show this warming.

A 2007 paper by Dr. David Douglass, physicist at the University of Rochester, and others, reviewed data from four independent weather balloon and three satellite data sets, and compared these with projections from 22 climate models. The paper concluded:

> On the whole, the evidence indicates that model trends in the troposphere are very likely inconsistent with observations…[38]

Dr. David Evans, a climate researcher from Australia, summarizes the powerful impact of this result:

> Weather balloons had been measuring the atmosphere since the 1960s, many thousands of them every year. The climate models all predict that as the planet warms, a hot spot of moist air will develop over the tropics about 10 kilometres up, as the layer of moist air expands upwards into the cool, dry air above. During the warming of the late 1970s, '80s and '90s, the weather balloons found no hot spot. None at all. Not even a small one. This evidence proves that the climate models are fundamentally flawed, that they greatly overestimate the temperature increases due to carbon dioxide.[39]

So how did the proponents of man-made warming react to this data? *They concluded that the models must be right and that the seven balloon and satellite data sets must be wrong.* Climatists had expected empirical data to back up the models, but by the mid-1990s it became clear that weather balloon and satellite measurements did not show the expected hot spot.

For the last fifteen years, the advocates of man-made warming have been *desperate* to show that the empirical data is wrong. More than one hundred papers have been written to point out the "uncertainties" in the weather balloon data, claiming problems such as "time-varying biases" and "temporal inhomogeneity."

Garbage In, Garbage Out

"People underestimate the power of models. Observational evidence is not very useful."
—John Mitchell,
Principal Research Scientist,
UK Meteorological Office[40]

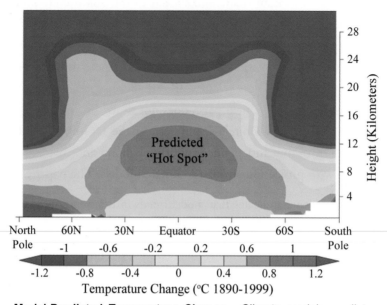

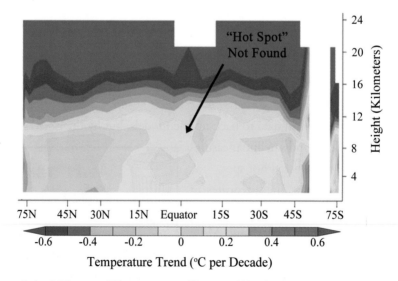

Model-Predicted Temperature Change. Climate models predict a "hot spot" in the troposphere centered over the Equator. (Adapted from IPCC, 2007)[41]

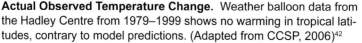

Actual Observed Temperature Change. Weather balloon data from the Hadley Centre from 1979–1999 shows no warming in tropical latitudes, contrary to model predictions. (Adapted from CCSP, 2006)[42]

Two in-depth expert panel assessments were conducted to try to explain why model projections did not match measured data. A paper by Allen and Sherwood in 2007 even claimed that using wind speed data to measure temperature better matched model projections than weather balloon temperature sensors.[43] But today, no empirical data exists that supports the model prediction of a hot spot in the troposphere over the tropics.

CONCLUSION #5: IGNORE THE SUN?

The IPCC states:

> …we conclude that it is very likely that greenhouse gases caused more global warming over the last 50 years than changes in solar irradiance [sunlight]."[44]

Thus the IPCC concludes that carbon dioxide, a trace gas in our atmosphere, has more influence on our climate than the sun. They discount the sun, the massive star that provides more than 99 percent of the energy from space absorbed by our Earth. This is the same sun that, by uneven heating of the tropics and the poles, drives all weather on Earth.

The fifth false conclusion of Climatism is that the sun is an insignificant factor in climate change. Prior to the launch of satellites in 1979, historical measurements had not been able to detect any change in the amount of solar radiation received by Earth. As a result, scientists believed for many years that the sun's energy output was a "solar constant," essentially unchanging. Today satellites measure the radiation received at the top of Earth's atmosphere at about 1,366 watts per square meter, with a small variation of about 0.1 percent between the peaks and valleys of each eleven-year solar cycle.[45] Since this variation is too small to account for the 0.5°C temperature rise from 1975 to 1998, the IPCC has ruled out the sun's influence and concluded that man-made greenhouse gases are driving climate change. But it appears that the IPCC jumped to an early conclusion in discounting the sun's impact.

There is good evidence that the climate impacts of the sun *do* change over time. Although we don't have good historical measurements of solar output levels, we do have historical records of sunspot activity. Sunspots are dark spots that appear on the surface of the sun that can easily be seen through a telescope. They are areas of intense magnetic activity that are often many times larger than Earth. Sunspots, solar flares, and coronal holes can be regarded as measures of solar activity. Sunspots have been tracked by astronomers since the 1600s.

In 1997, Douglas Hoyt and Kenneth Schatten published a reconstruction of sunspot activity from 1610 to 1995. The data showed periods of low activity, including the Maunder Minimum of 1650–1700, a period of almost complete lack of sunspots, and the Dalton Minimum at about 1815. The Maunder Minimum coincided with the coldest years of the Little Ice Age. The analysis also showed an increasing level of sunspot activity over the last 400 years that closely paralleled the rise in Earth surface temperatures from the depths of the Little Ice Age into today's Modern Warm Period.[46]

In an earlier paper in 1993, Hoyt and Schatten noted that:

> The correlation of the solar indices and modeled solar irradiance with the Earth's temperature are significant at better than the 99% confidence level.[47]

But the authors were puzzled that, although the correlation was very high, the changes in levels of irradiance (sunlight) were not large enough to explain the temperature changes, unless the climate was "much more sensitive to solar forcing than is commonly assumed."[48] Clearly, the sun was a factor in climate change, but something more than sunlight was at work—a missing link.

In addition to the sun's gravitational pull and the radiance of sunshine, the sun also bathes Earth in a stream of particles known as the solar wind. The solar wind is a continuous

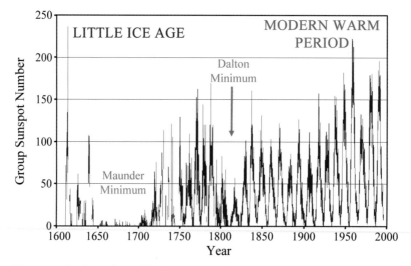

Sunspot Activity from 1610 to 1995. Sunspot cycles derived from the number of sunspot groups. The peaks shown are the maximum of each sunspot cycle every eleven years. Note the minimum activity during the Little Ice Age and the high activity during the Modern Warm Period. (Hoyt et al., 1998, adapted from *climatedata.info*)[49,50]

stream of high-speed electrons, protons, and other particles that travel outward from the sun. The solar wind is best known to cause the aurora borealis, also known as the northern lights. The aurora borealis are streamers of lights that appear in the sky in arctic regions. When present in the Antarctic, they are called aurora australis, or the southern lights. The auroras are caused when particles of the solar wind interact with Earth's magnetic field. When sunspots, solar flares, and other activity on the surface of the sun occur, the strength of the solar wind increases.

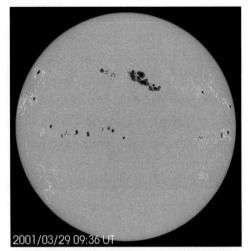

2001/03/29 09:36 UT

The Sun and Sunspots. Huge sunspot area on March 30, 2001. (SOHO, 2011)[51]

COSMIC RAYS AND CLOUDS—THE CLIMATE MISSING LINK

Scientists at the Danish Space Research Institute developed a breakthrough theory in the 1990s to explain the high correlation between solar activity and the climatic changes of Earth. They observed that the interaction of sunspots, the solar wind, and cosmic rays appeared to have a major impact on Earth's temperatures through cloud formation. This interaction was not considered by the IPCC, which only seriously considered variations in the level of solar radiation.

In addition to particles from the solar wind, Earth is also continuously under bombardment from cosmic rays. Cosmic rays are subatomic particles that shower Earth from deep space. Cosmic rays are created from stars that exploded ages ago in the Milky Way galaxy. They enter our atmosphere and collide with the atoms of atmospheric gases, forming secondary particles that reach Earth's surface. Earth's magnetic field provides a good barrier, shielding us from most cosmic rays.

The solar wind also acts to shield Earth from cosmic rays. Scientists noted decades ago

Aurora Borealis. Aurora over Bear Lake in Alaska, caused by interaction with the solar wind and Earth's magnetic field.
(USAF, 2006)[52]

that, when the level of sunspots is high, fewer particles are counted in cosmic ray monitoring stations. Many sunspots provide a strong solar wind, which blocks more of the cosmic rays aimed at Earth, reducing the number of cosmic ray particles that reach Earth's surface.

In 1995, Dr. Nigel Marsh and Dr. Henrik Svensmark of the Danish National Space Center published a paper that showed a correlation between the level of cosmic rays measured at Earth's surface and Earth's cloud cover. The data showed that when the level of cosmic rays increased, the level of low-altitude clouds increased over the 21-year period from 1984 to 2005.

The findings of Marsh and Svensmark are remarkable in two ways. First, the variation in cloud cover was found to be large, up to a two percent change over a single eleven-year sunspot cycle. This is about 20 times larger than measured changes in solar radiation, large enough to account for global surface temperature changes in the twentieth century. Second, the researchers found that the cosmic ray-to-cloud relationship was valid for low clouds, but not for middle and high-altitude clouds. A small change in low-altitude clouds can have a major effect on surface temperatures. Marsh and Svensmark postulated that a high level of sunspot activity reduces the level of cosmic rays and therefore reduces the level of low-level cloudiness, resulting in an increase in global temperatures.[53]

But, how can cosmic rays affect cloud cover? An admitted weakness of the Global Circulation Models is their inability to model clouds. This cosmic-ray-cloud link discovered by the Danes was completely unanticipated by the models. Most of the scientific community initially rejected the work of Marsh and Svensmark, because there was no mechanism known for cosmic rays to influence low-level clouds.

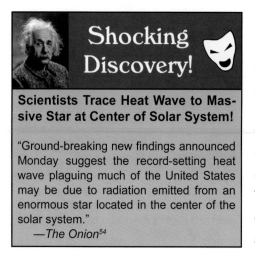

Shocking Discovery!

Scientists Trace Heat Wave to Massive Star at Center of Solar System!

"Ground-breaking new findings announced Monday suggest the record-setting heat wave plaguing much of the United States may be due to radiation emitted from an enormous star located in the center of the solar system."
—*The Onion*[54]

But soon afterward, Svensmark and his team in Copenhagen developed a theoretical basis and experimental data to support their theory. Named the SKY experiment, it consisted of a chamber filled with air, like our atmosphere, with controlled amounts of water vapor, sulphur dioxide, ozone, and other gases. Cosmic rays entered the chamber through the ceiling, as they bombard all of Earth's surface. The scientists found that when small amounts of sulfur dioxide were added to the chamber, sulfuric acid ions were created by collisions

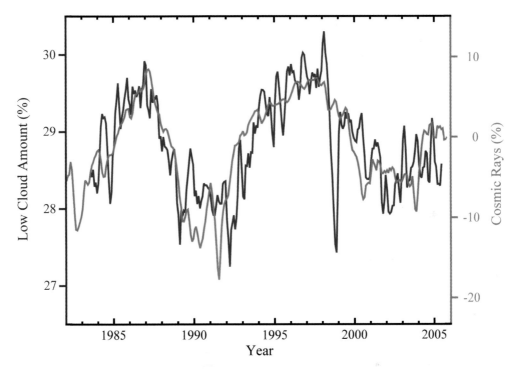

Variation in Low Clouds and Cosmic Rays 1984–2005. Cosmic ray variation closely tracks the global level and variation of low clouds according to satellite data from the International Satellite Cloud Climatology Project. (Adapted from Svensmark, 2007)[55]

with cosmic ray particles, resulting in large numbers of water droplets. Svensmark observed that the experiment provided a new mechanism for cloud formation:

> The theory describes mathematically the early growth of sulphuric acid droplets in the atmosphere. These are the building blocks for the cloud condensation nuclei on which water vapor condenses to make clouds.[56]

Meteorologists have long known that clouds must initially form on dust or aerosol particles. Such particles provide a platform on which water vapor molecules can condense to start the cloud formation process. Cloud seeding using salt crystals is a method that uses this principle. The SKY experiment showed that sulfuric acid ions created by cosmic rays are very effective condensation nuclei for forming clouds.

It appears that Svensmark and his team have found the climate "missing link." The high correlation of solar activity to Earth's temperature identified by Hoyt and Schatten, that was not able to be explained by small changes in the level of the sun's radiation, can be explained by the sun's effect on cosmic rays. Sunspot activity strengthens the solar wind, or solar magenetic field, which blocks a portion of the cosmic rays that enter Earth's atmosphere. Fewer cosmic rays means fewer created ions from collisions with atmospheric gases. Fewer ions provide fewer cloud condensation nuclei for the formation of low-altitude clouds. Less low-level cloudiness reflects less sunlight, so more sunlight is absorbed by Earth's land and oceans, causing Earth to warm.

Needless to say, Svensmark's theory was not well received by the man-made global warming scientific community. After Svensmark's first article about his theory was published in 1996, Bert Bolin, former Chairman of the IPCC, publicly pronounced Svensmark's theory as "scientifically extremely naive and irresponsible."[57] For the last 15 years, his theories have been heavily criticized by the Climatist community, including his book *The Chilling Stars*.[58]

After more than ten years of uncertainty, other experiments are now confirming the cosmic ray-cloud hypothesis of Svensmark. In May 2011, a second Danish experiment led by Dr. Ulrik Uggerhoj of Aarhus University used a particle accelerator to shoot electrons into a climate chamber in simulated atmospheric conditions. Cosmic ray collisions with atmospheric gases produce a shower of secondary particles, including electrons. The Aarhus results showed that the electron beam radiation stimulated the formation of aerosols, the condensation nuclei for cloud formation, supporting Svensmark's theory.[59]

Then in August 2011, the European Organization for Nuclear Research (CERN) reported new results from their CLOUD project in Switzerland. The CLOUD project (Cosmics Leaving Outdoor Droplets) was established as a follow-on effort to Svensmark's SKY experiment, using the big particle accelerator at CERN. Initial CLOUD experiments fired charged particles into a chamber containing a controlled gas mixture, simulating the effects of cosmic radiation in the atmosphere. The CERN press release reported that cosmic rays enhance the formation rate of aerosols by ten-fold or more. Project leader Dr. Jasper Kirkby reported:

> We've found that cosmic rays significantly enhance the formation of aerosol particles in the mid-troposphere and above.[60]

The CERN results were reported under constraining pressure from the global warming science community, because the results could seriously cripple the theory of man-made climate change. CERN Director General Rolf-Dieter Heuer directed that CLOUD scientists refrain from drawing any conclusions from the results:

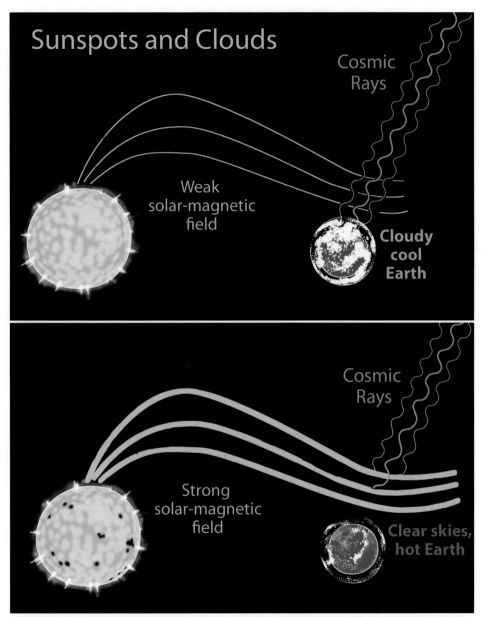

Sunspots and Global Warming. The level of sunspot activity affects low-level cloudiness and global temperatures. The bottom image shows global warming, where increased sunspot activity provides a stronger solar-magnetic field, reducing the level of cosmic rays that enter Earth's atmosphere. Fewer cosmic rays produce fewer cloud condensation nuclei, reducing low-level cloudiness. Reduced cloudiness reflects less sunlight, causing Earth to warm. The top image shows the opposite effect and global cooling. (Nova, 2011)[61]

I have asked the colleagues to present the results clearly, but not to interpret them…That would go immediately into the highly political arena of the climate change debate.[62]

Adding weight to Svensmark's theory is evidence that it appears to explain changes in global temperatures as we go back farther in history. Cosmic rays create Carbon 14 (^{14}C) as they bombard our planet. ^{14}C is "heavy carbon," with two extra neutrons in the nucleus of each carbon atom, the same atom used in carbon dating techniques. After being created in the atmosphere, ^{14}C accumulates in ice cores, cave stalagmites, trees, and other sites to give us a record of cosmic ray intensity throughout history. The more ^{14}C present in the environment, the higher the level of cosmic rays, and according to Svensmark's theory, the more low-level clouds, resulting in lower global temperatures. Therefore, high ^{14}C concentration means colder global temperatures.

Indeed, as we look back over the last 1,000 years, we find that an upside-down curve of Carbon 14 concentration tracks Earth's climate history very well. The Medieval Warm Period, the Little Ice Age, and the Modern Warm Period can be clearly seen, providing strong evidence for the sun as a driver of Earth's climate through the cosmic ray-cloud formation mechanism. The ice core proxies of atmospheric carbon dioxide show no such

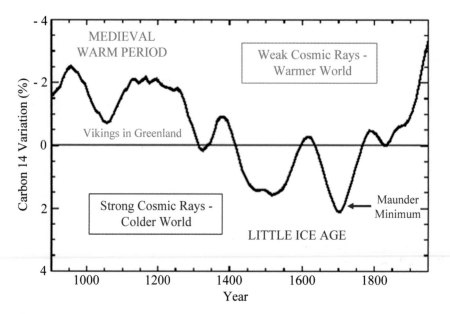

Carbon 14 over the Last 1,000 Years. Carbon 14 variation is plotted upside down to match temperature changes over the last millennium. ^{14}C is a well-known proxy for cosmic rays but also shows the Medieval Warm Period, the Little Ice Age, and the Modern Warm Period. (Adapted from Svensmark, 1999)[63]

1,000-year correlation with global temperatures.

Scientific evidence that the sun is our climate driver is increasing. Solar activity better explains both twentieth century and millennial temperature variations than does atmospheric carbon dioxide. Svenmark's theory that climate change is due to solar-driven changes to cosmic ray levels, resulting in variation in low-level cloudiness, is being confirmed by experiment.

THE WORLD *IS* AS IT SEEMS

In summary, the world is as your senses tell you, but not as you have been told. When you walk outside you feel the heat of the sun's rays, or the moisture of the rain, or the cooling effect of the clouds. If you live by the sea you can see the raging power of the ocean. The sun, weather and clouds, and ocean cycles are the dominant forces that shape Earth's climate. Carbon dioxide, that invisible trace gas that is blamed for our predicted climate destruction, is only a tiny part of the picture.

CHAPTER 6

THE OCEANS ARE RISING!
THE OCEANS ARE RISING!

"Never has good weather felt so bad. Never have flowers inspired so much fear.
Never has the warm caress of a sunbeam seemed so ominous.
The weather is sublime, it's glorious, it's the end of the world."
—JOEL ACHENBACH, STAFF WRITER, WASHINGTON POST, JAN. 7, 2007

Rising ocean levels are one of the greatest pending catastrophes proclaimed by Climatism. Warming global temperatures, resulting in the melting of Earth's icecaps, are the predicted cause. The 2006 Stern Report stated:

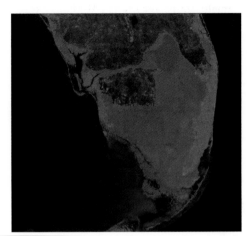

South Florida Flooding from Sea-Level Rise. Red areas show flooding from a simulated sea-level rise of five meters. (Weiss and Overpeck, University of Arizona, 2011)[1]

Rising sea levels will result in tens to hundreds of millions more people flooded each year with a warming of 3 or 4°C. There will be serious risks and increasing pressures for coastal protection in South East Asia (Bangladesh and Vietnam), small islands in the Caribbean and the Pacific, and large coastal cities, such as Tokyo, Shanghai, Hong Kong, Mumbai, Calcutta, Karachi, Buenos Aires, St. Petersburg, New York, Miami and London.[2]

Al Gore is a leading siren for sea-level calamity. In the book and movie versions of *An Inconvenient Truth*, he showed simulated color photographs of the flooded cities of Calcutta, New York, Beijing, and San Francisco, and vast areas of South Florida, Bangladesh, and the Netherlands. Gore warned of a 20-foot sea-level rise that would force millions to flee and cause billions of dollars of damage.[3]

James Hansen receives the award for the most alarming forecast. In 2006, he stated that the Greenland icecap was melting "far faster than scientists had feared." He warned:

> The last time the world was three degrees warmer than today—which is what we expect later this century—sea levels were 25m [meters] higher. So that is what we can look forward to if we don't act soon.[4]

Sea-level rise in the twenty-first century, if measured in meters, would certainly be a disaster for coastal cities and nations around the world. But what can we really expect? Are the icecaps of Earth melting at an alarming rate? In fact, the data shows that Earth's icecaps are in no imminent danger of melting. Let's examine the evidence.

EVIDENCE ABOUT EARTH'S ICECAPS

In the fall of 2007, the arctic icecap reached a 30-year low, according to satellite data. Many climate scientists issued forecasts predicting the near-term disappearance of arctic ice. Professor Wieslaw Maslowski of the Naval Postgraduate School in Monterey California projected that summer ice would be gone by 2013.[5] Professor Peter Wadhams of Cambridge University stated:

In the end, it will just melt away quite suddenly. It might not be as early as 2013 but it will be soon, much earlier than 2040.[6]

Both *BBC News* and the *New York Times* ran feature stories on the low level of ice. Andy Revkin of the *New York Times* used many alarming phrases in his article:

> Scientists are also unnerved by the summer's implications for the future…proof that human activities are propelling a slide toward climate calamity…humans may have tipped the balance…a particularly harsh jolt to polar bears.[7]

Before we jump to any conclusions, there are several aspects about arctic ice that we need to discuss. First, arctic ice tends to come and go, driven by natural cycles, and there is evidence that similar low levels of arctic ice occurred many times in Earth's climate history. While our satellite data goes back only 30 years, historical records describe past variations in arctic ice. George Ifft, the American consul at Bergen, Norway, sent a report on October 10, 1922, titled "The Changing Arctic" to the US State Department in Washington. The

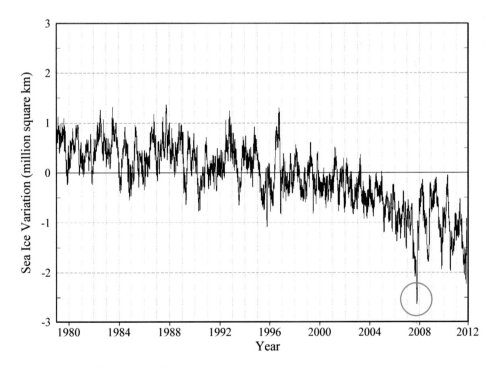

Arctic Sea Ice 1979–2011. The graph shows change in arctic sea ice from a mean value over the last 30 years. The ice reached a 30-year minimum in the fall of 2007. Data from satellites. (University of Illinois, 2011)[8]

report described an expedition to Spitzbergen and Bear Island, far north of Norway, and the lack of arctic ice in the year 1922:

> Ice conditions were exceptional. In fact, so little ice has never before been noted. The expedition all but established a record, sailing as far north as 81°29′ in ice-free water. This is the farthest north ever reached with modern oceanographic apparatus.[9]

Second, it's important to know that arctic ice floats in the Arctic Ocean. Should arctic sea ice melt entirely, whatever the cause, it would not raise global sea level at all. Floating ice does not raise the level of water by melting. You can demonstrate this with ice in a glass of water at home. For ocean rise to occur, the land-based icecaps of Greenland or Antarctica would need to melt.

Third, arctic ice is only about one to two percent of global ice. Climatists claim the melting of arctic ice is evidence that Earth's icecaps are melting. If someone says, "The elephant's trunk is in the water, so the elephant must be in the water," the next question asked should be, "What about the rest of the elephant?"

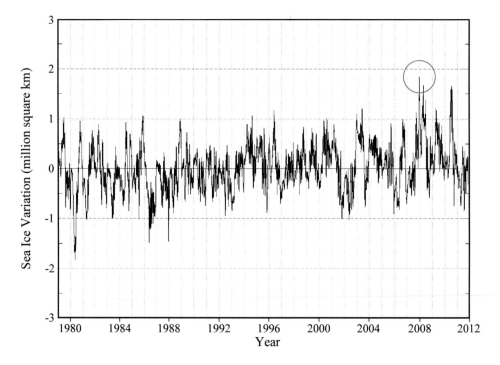

Antarctic Sea Ice 1979–2011. The graph shows change in antarctic sea ice from a mean value over the last 30 years. Note the 30-year maximum during the fall of 2007. Data from satellites. (University of Illinois, 2011)[10]

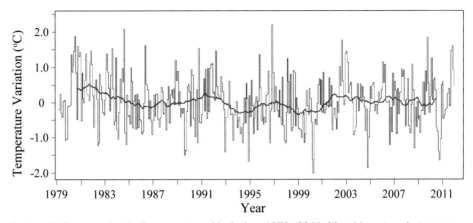

Antarctic Tropospheric Temperature Variation 1979–2011. Monthly antarctic temperature change data from MSU UAH satellites shows flat to declining temperatures over the last 30 years. The thick line is a 37-month running average. (Data from Christy and Spencer, University of Alabama Huntsville, graph adapted from *Climate4You*, 2011)[11]

Most of the rest of the elephant is the antarctic icecap, which accounts for 90 percent of Earth's ice. And lo and behold, not two weeks after arctic sea ice reached a 30-year minimum in the fall of 2007, *antarctic sea ice reached a 30-year maximum!* Satellite observation shows that antarctic sea ice has been rising for the last 30 years, in the opposite direction of the arctic trend. Yet the silence about the "good news" of growing antarctic ice has been deafening. The *BBC*, *New York Times*, and most other big names of the news media failed to report that antarctic ice is growing, a trend contrary to global warming dogma.

While satellite measurements show a warming Arctic over the last 30 years, they show a net cooling of Antarctica. Central and Eastern Antarctica, which contains about 90 percent of the ice mass of the continent, has been flat to cooling over at least the last 35 years. The Antarctic Peninsula in the west shows a moderate warming of 0.2 to 0.4°C.[12]

The Global Circulation Models are stumped when it comes to predicting polar temperatures. According to the models,

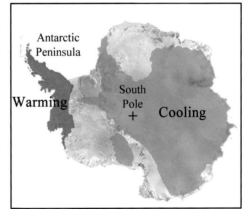

Temperature Trends in Antarctica. Temperatures for most of the Antarctic continent are flat to cooling over the past 35 years. The Antarctic Peninsula shows a moderate warming. (Adapted from Deyo, copyright University Corporation for Atmospheric Research 2008)[13]

Amundsen-Scott station at the South Pole. The station sits on stilts and can be jacked up over accumulating snow.[14]

temperatures should increase faster at the poles than other regions due to greenhouse gas-driven warming. But arctic temperatures are not rising faster than other regions, and antarctic temperatures are not rising at all.

By every indication, the antarctic icecap is healthy. Hands-on evidence comes from the Amundsen-Scott station at the South Pole. The United States has manned a South Pole station continuously since 1956. The station at the pole has been demolished, rebuilt, and expanded several times over the last 55 years.

A significant challenge with maintaining a South Pole station is accumulating ice and snow. The first two stations erected in 1956 and 1975 were eventually buried by mounting snowfall. The South Pole receives an incremental seven to eight inches of snow every year that continues to pile up year after year. The current station is a modular design, with the ability to be jacked up to a higher elevation, to maximize its useful life.[15]

Greenland Moulin. (Photo by Roger Braithwaite, University of Manchester)[16]

The Greenland icecap contains about eight to nine percent of global ice, with the last one percent held in Earth's mountain glaciers. Greenland is a favorite for alarming global warming documentaries on the *Public Broadcasting Station* and other television channels. Video of melting ice water pouring into a crevasse is shown as evidence of abnormal changes to the Greenland icecap. But melting ice on the continent of Greenland is a normal event. Greenland average annual temperatures vary around 0°C, the freezing point of water, so winter freezes and summer melts are typical. A look at historical temperatures for Angmanssalik and Gotthab Nuk, the two stations with the longest running data sets, show that on average, Greenland temperatures were warmer

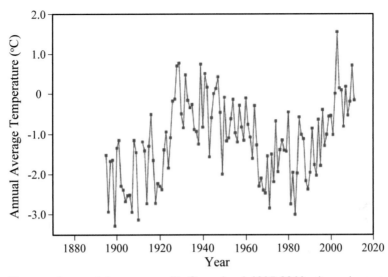

Temperatures at Angmagssalik Greenland 1895-2011. Annual mean temperatures at Angmagssalik, located on the southeast coast of Greenland (65.6 N, 37.6 W). Even with the warm spike in 2003, average temperatures for the 1930s–1940s were higher than for the 1990s–2010s. (Adapted from NASA GISS, 2011)[17]

during the decades of the 1930s and 1940s than the decades of the 1990s and 2000s. The graph on page 63 of Chapter 4 was based on ice cores from a site at Crete, Greenland. The data shows that Greenland was warmer many times in the past than it is today.

Evidence for a healthy Greenland Icecap is found in the story of the P-38 aircraft nicknamed the Glacier Girl. The Glacier Girl was one of eight aircraft on the way from the United States to England by way of Greenland in 1942. The planes got caught in a storm and were forced to crash land on the Greenland ice sheet. After nine days on the ice, all 25 aircraft crewmen were rescued by the US Coast Guard and the planes were abandoned.

In 1981, businessmen Pat Epps and Richard Taylor formed the Greenland Expedition Society and mounted an effort to recover the planes. Taylor recalls:

Our thoughts were that the tails would be sticking out of the snow. We'd sweep off the wings and shovel them out a little bit, crank the planes up and fly them home. Of course, it didn't happen.[18]

Despite the fact that the crash site location

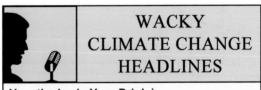

WACKY CLIMATE CHANGE HEADLINES

How the Ice in Your Drink is Imperiling the Planet
—Jeffrey Kluger, senior editor, *TIME*, Apr. 14, 2011[19]

was well known, the planes were nowhere to be found. Three more expeditions were mounted in the early 1980s, but were also unable to locate the planes.

In 1988, Epps and Taylor returned for a fifth time, this time equipped with sophisticated sub-surface radar. They quickly found all eight planes within two miles of the original crash location. *The radar found the planes buried under 270 feet of solid ice!* In 46 years, 270 feet of ice and snow had accumulated on top of the planes, on the supposedly "melting icecap of Greenland." The expedition bore a vertical shaft to reach the Glacier Girl. Over a several-year period the plane was disassembled, brought to the surface, and re-assembled and restored to flight worthiness in the US.[20]

In fact, the Greenland icecap is stable. It's melting around the edges but accumulating ice and snow in the center. Temperature and historical data show that neither the antarctic icecap nor the Greenland icecap is in near-term danger of melting, and that the current small size of the arctic icecap is not abnormal in Earth's history.

SEA-LEVEL RISE—WHAT WE CAN REALLY EXPECT

So what does all this mean for ocean levels? Oceans have been rising for thousands of years. Since the last ice age 20,000 years ago, sea levels have risen about 120 meters (390 feet) according to NASA.

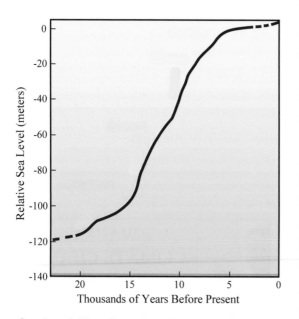

Over the last 150 years, oceans have been rising at the rate of about seven to eight inches per century. This rate has varied from zero and twelve inches per century for short periods of time. Man's heavy emissions of carbon dioxide really began the 1940s. But scientists have not measured an acceleration in sea rise during the last 50 years despite an increasing level of emissions. With the Greenland icecap stable and the antarctic icecap growing, a 20-foot rise in sea levels just isn't going to happen. Expect a 7–8 inch rise over the next century.

Sea-Level Rise Since Last Ice Age. Sea levels have risen about 120 meters (390 feet) in the last 20,000 years. (Adapted from NASA, 2007)[21]

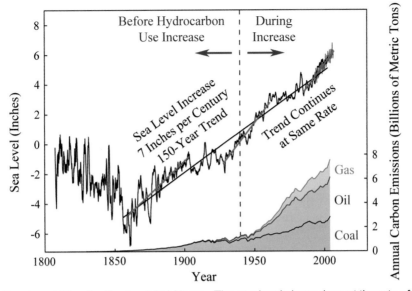

Sea-Level Rise for the Last 150 Years. The sea levels have risen at the rate of about 7 inches (18 cm) per century over the last 150 years. Intermediate trends are between 0 and 12 inches per century, but the rate has not accelerated after the large human hydrocarbon use. Data is from tide gauges from 1807 to 2002 and from satellites from 1993 to 2006. (Robinson et al., 2009)[22]

PREPARATIONS: FOUNDED AND UNFOUNDED

The Republic of Maldives is an island nation located in the northern Indian Ocean, 250 miles (400 kilometers) southwest of India. More than 300,000 people live on 26 atolls at an average height of 1.5 meters (4.9 feet) above sea level. Male, the nation's capital, is sheltered by sea walls, but most of the rest of the country has no protection. The residents of Maldives are understandably concerned about rising seas.

In 2009, the President Mohamed Nasheed and his cabinet used scuba gear to meet underwater to sign a declaration calling for global cuts in carbon emissions at the upcoming United Nations climate summit in Copenhagen. President Nasheed stated:

> We are trying to send our message to let the world know what is happening and what will happen to the Maldives if climate change isn't checked.[23]

Maldives' plight is serious, because rise in ocean levels from natural causes is likely to continue. But the idea that we can halt this rise by stopping industrial emissions of carbon dioxide, which are only about three percent of the natural CO_2 emissions from land and oceans, is nonsense.

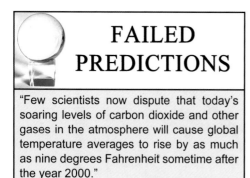

FAILED PREDICTIONS

"Few scientists now dispute that today's soaring levels of carbon dioxide and other gases in the atmosphere will cause global temperature averages to rise by as much as nine degrees Fahrenheit sometime after the year 2000."
—Carl Sagan, *The Vindicator*, Dec. 12, 1985[24]

Many communities across the United States are also concerned about sea-level rise and have deployed wide-ranging programs to "fight" climate change. Broward County, Florida, is an example. Broward is a coastal county in southeast Florida, containing Fort Lauderdale, the largest city in the county. The county issued a 113-page report in 2010, titled "Climate Change Action Plan" that is similar to plans in many communities across the US.[25]

The Broward plan projects an ocean rise of 24–48 inches by the year 2100. Although these levels are three to six times higher than the historical rise of 7–8 inches per century over the last 150 years, the plan rightly calls for adaptation by county communities. These measures include beach erosion control, standards for building construction, transportation adaptation, and protection of drinking water resources. But the plan also calls for a host of mitigation measures to "slow the impacts of climate change," which will have negligible effect on the climate. Efforts are underway to reduce the "carbon footprint" of county utilities; to establish building construction codes to reduce the use of water, electricity, and energy; to create a mass transportation system to "reduce vehicle emissions"; to plant trees to "store high levels of carbon"; and to promote wind, solar, and other renewable energy.[26] Broward is wasting large sums in a fantasy battle against the climate.

The San Francisco Bay Conservation and Development Commission is preparing for a 55-inch (140-centimeter) ocean rise by the year 2100. This is much larger than the average estimate of fifteen inches (38.5 centimeters) in the IPCC 2007 Fourth Assessment Report. Adaptation preparations are great, but a 55-inch rise is only based on model projections, not any historical trends or evidence.[27]

LET'S GO PLAY AT THE MELTING NORTH POLE

News reports that arctic ice is melting have inspired a comical group of explorers to try to reach this new "balmy" North Pole. In 2007, American Ann Bancroft and Norwegian Liv Arnesen set off on an unsupported trek to the North Pole. Bancroft was the first woman to reach the North Pole by foot and sled in 1986. According to Ann, the purpose of the 2007 trip was "to teach kids about global warming." But just eighteen miles into a 530-mile expedition, the two encountered temperatures of -60°C (-76°F) and had to be airlifted to safety. Bancroft suffered the loss of part of her big toe from frostbite.[28] You think she'll teach that to the kids?

"The consequences would be catastrophic. Even with a small sea-level rise, we're going to destroy whole nations and their cultures that have existed for thousands of years."
—Jonathan Overpeck,
    IPCC Lead Author, 2004[30]

The next year, British swimmer and ecologist Lewis Gordon Pugh set out from the northern island of Spitzbergen to kayak 1,200 km to the North Pole. His journey was intended to raise awareness about the melting arctic ice. But he was forced to stop after only 135 kilometers (km), blocked by arctic sea ice.[29]

In 2010, Tom Smitheringale, an adventurer from Western Australia, set out to be the third single explorer to reach the pole. His website said:

> Part of the reason Tom's One Man Epic is taking place now is because of the effect that global warming is having on the polar ice caps…Some scientists have even estimated that the polar ice cap will have entirely melted away by 2014![31]

Still 1,000 km from the pole, Smitheringale fell through the ice, suffered hypothermia, and almost died, before being rescued by the Canadian military.

THE SPECTER OF SEA-LEVEL RISE

Garbage In, Garbage Out

The specter of millions of the world's poor in flight from rising, crashing seas is a powerful vision. Many a world citizen has swallowed this apocalyptic image. But the antarctic icecap, 90 percent of the world's ice, is growing. The Greenland icecap is stable. Arctic sea ice reached a 30-year low in 2007, but arctic ice has been as low in times past. Sea levels are rising at seven to eight inches per century, driven

"The presenter was incredulous and asked Folland to repeat his statement so that the entire audience could hear, and Folland again said, 'The data don't matter…we're not basing our recommendations upon the data; we're basing them upon the climate models.'"
—Chris Folland, UK Meteorological
    Office (1991)[32]

Despite the claims of some climate scientists, Earth doesn't have one of these!

by natural causes, and a 20-foot rise is highly unlikely.

But, when did mankind get the notion that we could stop the rise of the seas? How did some get the idea that the oceans could be tamed by eliminating one CO_2 molecule from every 10,000 air molecules? As Dr. Willie Soon of Harvard University points out, "a magical CO_2 knob for controlling weather and climate simply does not exist!"[33] Adaptation to rising seas is the only real alternative for the Maldives and coastal cities. Sea levels will rise and fall from natural causes, and mankind will need to continue to adapt to climate change, just as we have done for thousands of years.

CHAPTER 7

WILD WEATHER AND SNOW FOLLIES

"I'm leaving because the weather is too good. I hate London when it's not raining."
—COMEDIAN GROUCHO MARX

Disaster struck a concert at the Indiana State Fair in Indianapolis on the night of August 13, 2011. At 8:49 p.m., unexpected wind gusts of up to 70 mph toppled a stage, killing five and injuring dozens of others. Just ten minutes before, the National Weather Service had issued a severe thunderstorm warning, but fair officials were still considering cancelling the concert when the stage suddenly collapsed.[1]

On Monday, August 18, ABC World News anchor Diane Sawyer announced "Weather Gone Wild" as the lead story for the evening news:

DIANE SAWYER: And tonight, the weather gone wild. Winds that come out of nowhere. Floods swelling streets. Heat breaking records in all 50 states. Snow where it hasn't fallen in decades. Something strange going on around the globe.

JIM AVILA: From the mid-Atlantic to New England, buckets of rain, a record ten inches fell on New York's Long Island yesterday...Never had anything like this heat either. Triple digits across Texas again today. Halfway through August, 5,000 heat records have been broken across the country. Every state in the US set a heat record, all 50. Waco hit 100 for the 63rd time this year, tying an all-time record. It was nature from another angle in Indianapolis over the weekend, straight-line winds, unseen on radar, out of nowhere, hit 70 miles per hour, knocked down the concert stage, killing five.[2]

People are obsessed with the weather. It's discussed in elevators and around the water cooler at work almost every day. We talk about it on telephone calls with distant family members. We drive through it, walk through it, and enjoy the great outdoors with it and in spite of it. Weather enables skiing and sailing, but forces cancellation of baseball, cricket, tennis, and other sporting events. A local storm or heat wave makes weather the lead story on our television channels. A hurricane approaching the coastal US produces round-the-clock coverage on major TV networks for days, as Hurricane Irene did in August 2011.

Of course, weather impacts major industries such as agriculture, construction, energy, fishing, retailing, and tourism. Tropical storm or tornado events heavily impact the insurance industry. Our food prices and even the health of national economies are impacted by crop yield and weather cycles.

Weather varies widely around the world. One summer, when people of Chicago were sweating in 100°F heat, my daughter reported from London that she could see her breath outside. Meteorology is a boring profession in Singapore, a city located 85 miles north of the equator. Daily high temperatures in Singapore reach a humid 87°F (31°C) day-after-day, almost every day of the year.[3]

But for most of the globe, weather is always changing. In Chicago, temperatures typically swing 100°F (56°C) during a normal year. The warmest summer day is usually about 95°F and the coldest winter day is about -5°F. Compare this to the increase in global surface temperatures of only 1.3°F since the American Civil War.

WISDOM FROM CLIMATISM?

"...man is now in charge of the thermostat for the globe."
—US Senator Tom Udall (2011)[4]

During every week of the year, somewhere on Earth, a drought, flood, storm, or forest fire occurs that appears to be a once-in-five-hundred-year event. These are a normal part of Earth's chaotic climate. But climate sirens now claim these events are due to global warming, and paradoxically caused by relatively tiny human contributions to a trace gas in our atmosphere. They've become a new class of weather alarmists.

THE WEATHER SCAREMONGERS

"Irene's got a middle name, and it's Global Warming." So said Bill McKibben in August of 2011 as Hurricane Irene approached Cape Hatteras, North Carolina. McKibben, Al Gore, and others have become weather scaremongers, alarmists who use any weather event to declare that humans are destroying Earth's climate and that we need to curb our evil carbon-emitting ways. McKibben went on:

> Remember…this year has already seen more billion-dollar weather-related disasters than any year in US history. Last year was the warmest ever recorded on planet Earth. Arctic sea ice is near all-time record lows. Record floods from Pakistan to Queensland to the Mississippi basin; record drought from the steppes of Russia to the plains of Texas…This is what climate change looks like in its early stages.[5]

McKibben went on to demand that President Obama deny approval of the Keystone Pipeline, a project to bring oil to the US from Canadian tar sands.

The spring of 2011 brought an outbreak of tornados across the southern United States. Five separate outbreaks totaling 1,000 tornados ripped across the south and central US, killing more than 500 people and inflicting $9 billion in damage.[6] A wet April in the Midwest and runoff from heavy winter snows forced the Mississippi River to overflow its banks, flooding millions of acres. Judi Greenwald of the Pew Center on Climate Change warned:

You can no longer say that the climate of the future is going to be like the climate of today, let alone yesterday.[7]

So, the tornados and the flooding were due to those four carbon dioxide molecules in every 10,000 of our atmosphere?

"When the weather's strange in your neighborhood, who you gonna call?"

WEATHER SCAREMONGERS

"WEATHER PANIC! THIS IS THE NEW NORMAL (AND WE'RE HOPELESSLY UNPREPARED)"
—Sharon Begley, *Newsweek* cover story, June 6, 2011[8]

Meteorologist Joseph D'Aleo predicted the tornado outbreaks several weeks in advance and also provided the reason for the spring floods of 2011. In early April, he pointed out that above-normal snow cover remained on the ground across the northern US and Canada. This kept the upper US cooler than normal, in sharp contrast to a warming Gulf of Mexico. In addition, the Pacific Ocean remained in a La Niña phase, driving a strong jet stream across the central United States.[9] The combination of La Niña winds and the temperature contrast between the cold north and the warm Gulf of Mexico produced the severe weather. Melting snow raised the flood waters. Yet, the Climatists ignored these natural explanatory factors and blamed the severe weather on man-made influences.

OF HURRICANES AND TORNADOS

In June 2005, Dr. Kerry Emanuel of the Massachusetts Institute of Technology published a paper that became the basis for storm alarmism. Emanuel examined tropical cyclones in both the Atlantic and Pacific Oceans from the years 1930 to 2005 and found that the frequency of storms was the same, but that the "potential destructiveness of hurricanes… has increased markedly since the mid-1970s." He claimed that increasing sea-surface temperature was the cause of the more severe storms.[10]

On August 29, 2005, hurricane Katrina hit New Orleans with Category 3-force winds estimated at 125 mph. The subsequent failure of levees in New Orleans made Katrina the most destructive US hurricane in history in terms of economic losses. That same year, hurricanes Dennis and Rita hammered the Gulf Coast, while Wilma devastated South Florida.[11] The years 2004–2005 rivaled 1993–1994 and 1997–1998 for the most active hurricane periods in the last 30 years.

Despite many opposing points of view, the IPCC concluded that global warming was causing stronger tropical storms in its 2007 Fourth Assessment Report, based on the Emanuel paper. The report stated:

"Heavy rains, deep snowfalls, monster floods and killing droughts are signs of a 'new normal' of extreme US weather events fueled by climate change, scientists and government planners said on Wednesday. 'It's a new normal and I really do think that global weirding is the best way to describe what we're seeing,' climate scientist Katharine Hayhoe of Texas Tech University told reporters." —*Reuters*, May 18, 2011[12]

"…it is *likely* that future tropical cyclones (typhoons and hurricanes) will become more intense, with larger peak wind speeds and more heavy precipitation associated with ongoing increases of tropical sea surface temperatures."[13]

Many scientists criticized

Dr. Emanuel's conclusions and the IPCC's position. Dr. Christopher Landsea, research meteorologist for the National Oceanic and Atmospheric Administration's National Hurricane Center in Orlando, Florida, pointed out that the historical record for both the number and strength of tropical storms is inaccurate. Much of Landsea's work has involved reconstruction of databases for tropical storms. He said:

"*Protecting* dozens of major coastal cities from future flooding will be challenging enough—*rebuilding* major coastal cities destroyed by super-hurricanes will be an almost impossible task."
—Author Joseph Romm, 2007[15]

> Trend analyses for extreme tropical cyclones are unreliable because of operational changes that have artificially resulted in more intense tropical cyclones being recorded, casting severe doubts on any such trend linkages to global warming.[14]

Dr. Landsea resigned from the IPCC in 2005, over the IPCC's alarming position on tropical storms forced by global warming, stating:

> I am withdrawing because I have come to view the part of the IPCC to which my expertise is relevant as having become politicized. In addition, when I have raised my concerns to the IPCC leadership, their response was simply to dismiss my concerns.[16]

Dr. William Gray, the renowned hurricane forecaster at Colorado State University, also levels serious criticism at the IPCC. Gray points out that sea-surface temperature is only one of many factors that impact the number and strength of tropical storms, and that empirical evidence does not show an increase in cyclone frequency or strength with increasing sea-surface temperature. He remarks:

Climate Books We Expect to See[17]

Hurricane Mitigation FOR Beginners

> The IPCC hierarchy had its mind made up years ago to make every attempt possible to link rising levels of CO_2 with increases in global hurricane intensity and frequency… Input from skeptics or any hypothesis or data that did not link rises in CO_2 to increases in tropical cyclone activity was to be avoided, suppressed, or rejected.

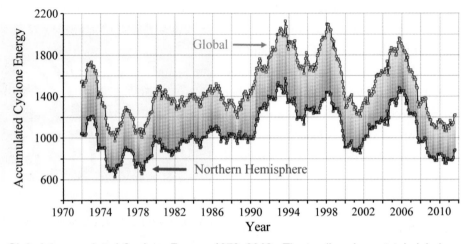

Global Accumulated Cyclone Energy 1972–2012. The top line shows total global cyclone energy for tropical storms and the bottom line shows the cyclone energy for the Northern Hemisphere. The ACE index is the sum of the squares of the maximum surface wind speed measured every six hours for all storms with wind speeds exceeding 35 knots. (Adapted from Maue, 2011)[18]

Gray goes on to conclude that "rising levels of CO_2 can have little or no significant influence on tropical cyclone frequency and intensity."[19]

So are hurricanes and tropical storms getting stronger? Satellites can now provide the answer to this question. Dr. Ryan Maue at Florida State University monitors global tropical storm activity on an around-the-clock basis. Wind scatterometers mounted on NASA satellites are able to look down and measure the wind speed of every tropical storm on Earth for each day of the life of the storm.[20] Dr. Maue compiles the measured maximum wind speed for all major storms into a metric called ACE (Accumulated Cyclone Energy). In 2005, when the Emanual paper was published and when Hurricane Katrina bashed New Orleans, global ACE did reach a 7-year peak. But in 2011, Maue reported that global tropical cyclone accumulated energy had fallen by half from 2005 and set a new record low, lower than the previous record in 1977.[21] So, Dr. Maue's ACE data shows *no trend of rising global tropical cyclone activity* over the last 35 years, despite steadily rising atmospheric carbon dioxide levels and generally increasing sea-surface temperatures since 1975.

Actually since 2005, US citizens have been fortunate to avoid any major devastating hurricanes. Hurricane Irene, which hit North Carolina and New York in 2011, was only a large tropical storm when it crossed the coastline in terms of its peak sustained wind speed. Over the last century, the US was hit by one Category 3 or stronger hurricane every two years. But not a single hurricane with Category 3 winds crossed the US coastline during 2005–2011. To find the last five-year period without landfall of a Category 3 hurricane, we need to go all the way back to 1910–1914.[22]

Tornado events in the US are now declared to be global warming events. Al Gore, Bill McKibben, Kevin Trenberth, and many others declared that the 2011 tornado outbreak was "influenced by," "affected by," "fueled by," or "contained fingerprints of" man-made global warming. Conclusive quotes from these scaremongers appear in the news after each weather event but are never accompanied by any data to show an increasing trend of stronger tornados. The reason is that *there isn't any such data.*

The National Climatic Data Center keeps an annual count of the number and estimated strength of tornados in the US. It is true that the total number of tornados recorded has been increasing. But this is due to better radar detection techniques and many of these funnel clouds never touch down. A graph of NCDC data shows that the number of strong tornados peaked in the 1970s and has been *declining* through 2010.

Dr. Roy Spencer points out that tornados do not require tropical-type warmth. Tornado outbreaks in the US are generally during cooler-weather years. Spencer remarks:

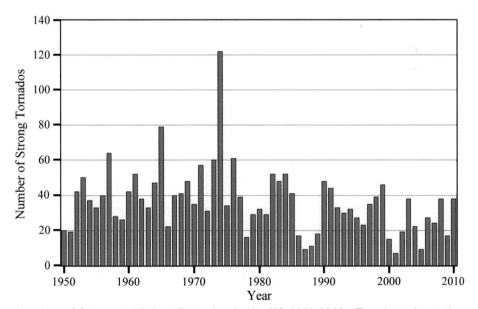

Number of Strong to Violent Tornados in the US 1950-2010. The chart shows the number of F3 to F5 tornados by year from 1950 to 2010, as measured on the Enhanced Fujita Damage Intensity Scale. The Fujita Scale measures tornado strength from F1 (weakest) to F5 (strongest). The numbers of strong tornados have been trending downward from the 1970s. (NCDC, 2011)[23]

It is well known that strong to violent tornado activity in the US has decreased markedly since statistics began in the 1950s, which has also been a period of average warming. So, if anything, global warming causes FEWER tornado outbreaks…not more. In other words, more violent tornados would, if anything, be a sign of "global cooling," not "global warming."[24]

Dr. Richard Lindzen's analysis agrees with Spencer. Lindzen reminds us that weather is caused by the temperature differential between the tropics and the polar regions. The larger this temperature difference, the greater the resulting storminess in Earth's mid-latitudes (in locations such as China, Europe, and the United States). From both climate change history and climate model predictions, we should expect greater warming in the polar regions than in the tropics. *This would reduce the temperature difference across the Earth, thereby reducing expected storminess.* Lindzen concludes:

> According to any textbook on dynamic meteorology, one may reasonably conclude that in a warmer world, extratropical storminess and weather variability will actually decrease.[25]

The claims of the alarmists are based on anecdotal events and short memories, without scientific basis. Neither data on global tropical cyclone activity nor US tornado data show a trend of increasing severe weather. Basic climate physics predicts reduced storminess as Earth warms. Yet, the weather scaremongers continue to alarm our citizens. Damn the facts, full speed ahead!

SNOW IS A SIGN OF GLOBAL WARMING?

For many years Climatism had been forecasting reduced snowfall due to global warming. In their book *The Climate Crisis,* David Archer and Steven Ramsdorf point to this conclusion from the IPCC Fourth Assessment Report in 2007:

> One of the robust findings of the report is that snow cover in most continental areas will dramatically decrease unless warming is stopped…Large areas are expected to become snow free.[26]

Dr. David Viner, senior research scientist at the Climatic Research Unit of the University of East Anglia, made the remarkable statement in 2000 that within a few years winter snowfall will become "a very rare and exciting event." He went on to predict that, "Children just aren't going to know what snow is."[27]

Scientists at the University of Wales tracked declining snowfalls throughout the 1990s and early 2000s. In 2007, they proclaimed that the 3,560 ft. Snowdon mountain may lose its snow cover by 2020. Lembit Opik, Minister of Parliament for Wales, lamented:

Snow-capped Snowdon has been an iconic Welsh image for centuries. It is shocking to think that in just 14 years, snow on this great mountain could become nothing but a permanent and distant memory.[28]

In the US, scientists were singing the same tune. Dr. Tim Barnett of the Scripps Institution of Oceanography published a paper in January 2008, finding that the

"…snow pack has decreased and been observed to melt earlier in the calendar year…the observed changes in the hydrological components…can be explained well by anthropogenic [man-made] forcing (green house gases and aerosols) alone.[29]

Gavin Schmidt and Joshua Wolfe agreed. On the cover of their 2009 book *Climate Change: Picturing the Science*, they used a photo of Lake Powell, a dammed reservoir on the Colorado River in Arizona and Utah.[30] The picture, taken after drought conditions in the West from 2000–2005, showed the lake at a very low level. The implication was that man-made climate change was causing the declining snowpack and western US drought conditions. On both sides of the Atlantic, Climatist scientists were pointing to reduced snowfall as evidence of climate change.

But then the weather changed. Britain was hit by three cold winters with increased snowfall from December 2008 through February 2011. Prior to each, the UK Meteorological Office used their multi-million pound sterling supercomputers and climate models to forecast milder winters, no doubt influenced by global warming dogma.

After the cold, snowy winter of 2008–2009, Met Office scientist Peter Stott declared the winter to be "an anomaly." He suggested that exceptionally cold winters were now expected to occur "about once every thousand years or more."[31] Alex Hill of the Met Office then upped the ante, declaring it "very unlikely" that there would be a skiing industry in Scotland in 50 years' time.[32]

Then in 2009–2010, the UK was socked with another abnormally cold and snowy season. Average temperatures for January and February in Scotland were the coldest since 1914. Northern Ireland experienced the coldest winter since the early 1960s, and England

Signs that Man-Made Warming is Happening???

British UFO Sightings at "Bizarre" Levels

"Some experts believe it could be linked to global warming and craft from outer space are appearing because they are concerned about what man is doing to this planet."
—*Telegraph*, July 7, 2008[33]

Met Office Follies

"Despite the cold winter this year, the trend to milder and wetter winters is expected to continue, with snow and frost becoming less of a feature in the future."
—Peter Stott, Met Office Scientist and IPCC Lead Author, Feb. 25, 2009[34]

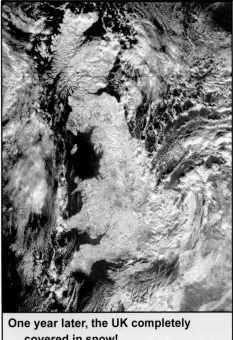

One year later, the UK completely covered in snow!
January 7, 2010 (NASA)[35]

and Wales were the coldest since the winter of 1981–1982.[36] On January 7, a spectacular satellite photo showed the entire British mainland covered in snow.

In July of 2010, Her Majesty's government held a "winter resilience review" to discuss the two recent harsh winters and plans for the upcoming winter season. David Quarmby of the Royal Automobile Club Foundation chaired the review. He recalls that the Met Office Hadley Centre climate research team advised that:

> …the chance of a severe winter in 2010/2011 is no greater (or less) than the current general probability of 1 in 20. The effect of climate change is to gradually but steadily reduce the probability of severe winters in the UK.[37]

Subsequently, the Met Office once again forecasted an "unusually dry and mild winter" as late as October 28, 2010, for the next upcoming season.[38] You guessed it, England experienced the coldest month of December since 1890.[39]

These successive failed predictions were so embarrassing that the Met Office *erased* them from their website. Heathrow Airport operator BAA, after cancelling 4,000 flights over five days during the December 2010 snowfall, announced in September 2011 that they had tripled their snow removal staff. Great idea to prepare for those future "mild and dry" winters.[40]

Snow was the norm for the winters of 2009–2010 and 2010–2011 around the world. On February 21, 2010, Moscow received more than 21 inches of snow in a single storm, with February exceeding the average by more than 50 percent. Just five months earlier, the mayor of Moscow promised that the Russian Air Force was now able to "keep it from snowing."[41] During the winter of 2010, major snowstorms buried New York City and

Chicago. Forty-nine of 50 states received snow (Hawaii the exception), a rare occurrence.[42] In 2010–2011, Northern Europe experienced an abnormally snowy and cold winter. Sweden experienced its coldest and snowiest winter in 100 years. Germany experienced its coldest December since 1969.[43] In the western US mountains, snow piled up during the winters of 2009–2010 and 2010–2011. By May of 2011, snowpack was two to three times normal in most western states.[44] By September 2011, the level of Lake Powell had risen to two and one-half times the average level, much higher than the doom-and-gloom cover picture of the Schmidt-Wolfe book.[45]

It is true that these are weather events, not climate, which is an average of weather events over periods of 30 years to centuries. But the 2008–2011 winters provided several "once-in-a-century" cold events in a row, each an unlikely occurrence according to climate model predictions. These many snowfalls were very embarrassing to man-made warmists, since snow was previously predicted to be "rare and exciting."

So what did the proponents of Climatism do? Just like the chameleon, *they changed their story*. Snowfalls now became "evidence" of global warming. Dr. Myles Allen, head of the Climate Dynamics group at the University of Oxford Department of Physics, stated:

> Even though this is quite a cold winter by recent standards, it is still perfectly consistent with predictions for global warming. As for snowfall, that could actually increase in the short term because of global warming.[46]

Come again? Or how about this statement in a 2011 paper from Daniel Huber and Dr. Jay Gulledge:

In December 2009 and February 2010, several American East Coast cities experienced back-to-back record-breaking snowfalls. These events were popularly dubbed "Snowmageddon" and "Snowpocalypse." Such events are consistent with the effects of global warming, which is expected to cause more heavy precipitation because of a greater amount of water vapor in the atmosphere.[47]

Signs that Man-Made Warming is Happening???

24 inches of snow, Washington D.C., Dec. 19, 2009[48]

Recall from Chapter 5 that global atmospheric water vapor has not detectably increased over the last 25 years, according to satellite data. So the conclusion from Huber and Gulledge is based only on computer model projections, not empirical data.

Of course, Mr. Gore agrees on the question of snowfall:

> As it turns out, the scientific community has been addressing this particular question for some time now and they say that increased heavy snowfalls are completely consistent with what they have been predicting as a consequence of man-made global warming.[49]

Well, there we have it. Snowfall is now evidence of anthropogenic global warming (AGW). So is lack of snow. So is rain, drought, flood, wildfire, hot weather, cold weather, sunshine during the day, starlight at night, and just about anything else you can think of. All of which caused UK journalist James Delingpole to quip: "Poo? In your woods? Then AGW definitely, truly exists!"[50]

THE GORE EFFECT

As the United Nations Climate Conference began in Cancún, Mexico, in early December 2010, Great Britain was lashed with snow and recorded the lowest temperature in 25 years. A frustrated George Monbiot, columnist for *The Guardian* and British global warming alarmist, used the headline, "Cancún climate change summit: Is God determined to prevent a deal?" Monbiot went on to lament:

> Is the divine presence a Republican? Or is He/She/It running an inter-galactic fossil fuel conglomerate?…whatever the explanation may be, the Paraclete appears to be as determined as any terrestrial corporate frontman to prevent a successful conclusion to the climate talks. How do I know? Because every time anyone gets together to try to prevent global climate breakdown, He swaths the rich, densely habited parts of the world with snow and ice, while leaving obscurer places to cook.[51]

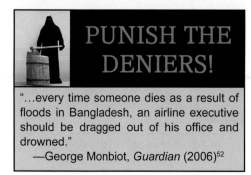

"…every time someone dies as a result of floods in Bangladesh, an airline executive should be dragged out of his office and drowned."
—George Monbiot, *Guardian* (2006)[52]

Mr. Monbiot was complaining about what has come to be known as "The Gore Effect." It seems that at almost every climate conference, public demonstration on global warming, and at many speeches by Mr. Gore himself, nature adds an ironic special touch of a snowstorm or record cold day. While Britain was beset by record cold and snow in December 2010, the

Cancún Climate Conference in Mexico experienced three record cold days itself.[53] The December 2009 Copenhagen Climate Conference was also blanketed by a snowstorm and frigid weather. Maybe God *is* a denier!

USE DISASTERS TO ENACT CLIMATE POLICY?

Climatists also *wish* for disasters so that the world can be convinced to take action to fight climate change. Economist Thomas Schelling is one of these:

> I sometimes wish we could have, over the next five or ten years, a lot of horrid things happening—you know, like tornadoes in the Midwest and so forth—that would get people very concerned about climate change.[54]

Sir John Houghton, while co-chairman of the IPCC, said, "If we want a good environmental policy in the future, we'll have to have a disaster."[55]

It's clear now that the governments of the world who accept the theory of man-made global warming also intend to use weather and natural disasters to persuade reluctant populations to adopt onerous climate policies. A 2011 report led by Sir John Beddington, Head of the UK Government Office of Science, recommended this course of action:

> The onset of more severe climate impacts overseas may also open up temporary opportunities, or "policy windows." These would allow legislators the licence to take specific bold actions which they ordinarily believe would not otherwise be possible or politically acceptable…In effect, envisaged solutions can become rapidly translated into practical options for action following a major disaster or near-miss.[56]

In other words, Sir John's team recommends the use of natural disasters to scare the British public so that unpopular climate change policies can be passed.

OUR MODERN AGE OF SUPERSTITION

The Aztecs of the 1500s practiced human sacrifice. Thousands died in gory rituals, including heart extraction, in which the priest would cut out the still-beating heart of the victim and hold it aloft to the honored god. The sacrifices were believed necessary to control the weather and keep the sun moving across the sky.

In the not-too-distant past of Renaissance Europe, persons accused of witchcraft were executed for weather-related "crimes." Dr. Emily Oster of Harvard University estimates that between the thirteenth and nineteenth centuries as many as one million people in Europe were executed for the crime of witchcraft. Dr. Oster states:

> The most active period of the witchcraft trials coincides with a period of lower than average temperature known to climatologists as the "little ice age"…In a time period when the reasons for changes in weather were largely a mystery, people would have searched for a scapegoat in the face of deadly changes in weather patterns. "Witches" became target for blame because there was an existing cultural framework that both allowed their persecution and suggested that they could control the weather.[57]

Are we so different today? Catastrophic warnings from computer models have replaced the oracles of the past. The new climate priests are Al Gore, Bill McKibben and others who take advantage of every storm to scare people. The new witches of our modern society are the coal and oil companies, which *must* be causing these deadly weather patterns. Only by destroying our industrial society can we calm the weather and stop the rise of the seas.

CHAPTER 8

BIG WHOPPERS
ABOUT CLIMATE CHANGE

"How many legs does a dog have if you call the tail a leg? Four.
Calling a tail a leg doesn't make it a leg."

—ABRAHAM LINCOLN

Climate science continues to be full of uncertainty. No one really knows the exact impact of man-made greenhouse gases on our planet. Despite the growing scientific evidence that this impact is small, claims of disaster grow ever more numerous and dire. Climate "whoppers" are delivered every day to citizens across the world. These whoppers are also taught to our children in school. How can trustworthy climate scientists let this propaganda go on? Let's discuss some of the biggest whoppers claimed by proponents of catastrophic global warming.

DANGEROUS CARBON POLLUTION?

Supporters of activist climate policies excel at developing effective media labels to promote their Earth-saving cause. Coal-fired power plants and oil refineries are branded as "dirty." Wind and solar facilities are labeled as "clean" and "free." Luxury cars are called "climate pigs" and electric vehicles are "eco-cars." "Climate change" itself is an effective media label, allowing any weather change to be blamed on man-made emissions. But one of the biggest whoppers is incessant hand-wringing about "carbon pollution" and a concern for "clean air."

"Dangerous carbon pollution" has become a standard phrase for environmental groups. For example, a May 2010 press release from the World Wildlife Fund called for "a science-based limit on dangerous carbon pollution that would send a strong signal to the private sector."[1] US public officials have adopted the phrase. Carol Browner, US Director of Energy and Climate Change policy, stated:

> The sooner the US puts a cap on our dangerous carbon pollution, the sooner we can create a new generation of clean energy jobs here in America…[2]

The fall of 2010 saw a battle raging in California. Opponents of the California Global Warming Solutions Act, Assembly Bill 32 (AB32), placed Proposition 23, a proposal to delay the act, on the ballot. More than $30 million was spent by opposing sides in the Proposition 23 fight. Prop 23 was defeated, a victory for climate nonsense, but the rhetoric in the campaign was surprising. Global warming advocates campaigned fiercely against Prop 23. They won by painting a false picture in the minds of California voters, claiming that AB32 was needed to reduce air pollution, reduce dirty energy, and to boost the creation of green jobs. They rarely even mentioned greenhouse gases!

December 7, 2009, is a date that will live in infamy. Not in memory of the attack on Pearl Harbor, but on that date the Environmental Protection Agency declared carbon dioxide a pollutant under the Clean Air Act. *That is bizarre*. Carbon dioxide is an invisible, odorless, colorless gas. It does not cause smoke or smog. A favorite image at the top of a typical

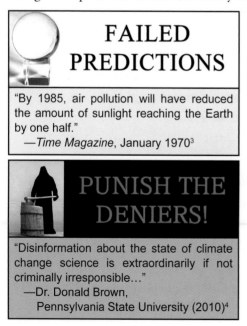

FAILED PREDICTIONS

"By 1985, air pollution will have reduced the amount of sunlight reaching the Earth by one half."
—*Time Magazine*, January 1970[3]

PUNISH THE DENIERS!

"Disinformation about the state of climate change science is extraordinarily if not criminally irresponsible…"
—Dr. Donald Brown,
Pennsylvania State University (2010)[4]

article about global warming shows a coal-fired power plant with a plume of white smoke trailing from the cooling tower. A similar picture graces the cover of Mr. Gore's *An Inconvenient Truth*. But that "smoke" isn't CO_2. It's condensing water vapor. We can't see CO_2.

Each person inhales only a trace of carbon dioxide from the atmosphere. But as our bodies burn sugars, we create carbon dioxide. We then exhale carbon dioxide at a level *100 times above that in the atmosphere* with every breath. Depending on room air conditioning, 50 people in a conference room can quickly

Ratcliffe Coal-Fired Power Plant in Nottinghamshire, UK. Rising plumes are condensing water vapor. CO_2 is an invisible gas. (Photo by Murray-Rust)[5]

produce enough carbon dioxide to raise room CO_2 levels to double the concentration in the atmosphere, a normal indoor condition. Atmospheric CO_2 is not harmful to humans until many times greater than current atmospheric levels.

It's true that incomplete combustion emits particles that can be harmful pollutants. I recall driving through Gary, Indiana, in the 1960s at the start of a family camping trip when I was young. As we passed the Gary steel mills, a thick cloud enveloped our station wagon, depositing a brown residue on our windshield. Air pollution was a serious problem in the United States 50 years ago.

But particulate carbon and other pollutants over US cities have been declining since the 1960s. Today, harmful pollutants are well-controlled by the Clean Air Act of 1970 and many other federal and state statutes. According to Environmental Protection Agency data, US air quality is much improved today, with six key air pollutants down a combined 57 percent since 1980. Since 1980, emissions of carbon monoxide are down 80 percent, lead down 93 percent, nitrogen dioxide down 48 percent, ozone down 30 percent, and sulfur dioxide down 76 percent. Carbon particulates have been tracked for fewer years, but large particulates are down 38 percent since 1990, and small particulates are down 27 percent since 2000.[6] This reduction has been accomplished at the same time that US citizens used 72 percent more electricity from coal-fired

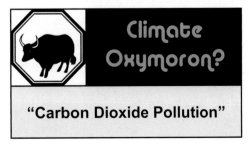

Climate Oxymoron?

"Carbon Dioxide Pollution"

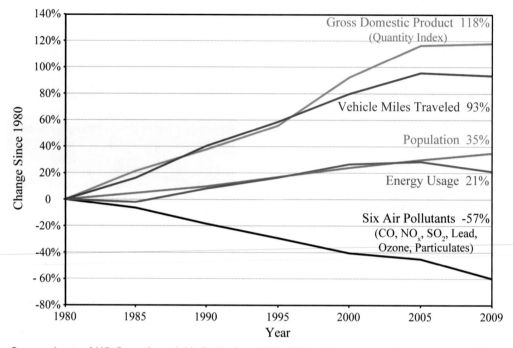

Comparison of US Growth and Air Pollution 1980–2009. A composite of six air pollutants has declined 57 percent from 1980 to 2009. At the same time, Gross Domestic Product is up 118 percent and vehicle miles traveled are up 93 percent. (EPA, US Census Bureau, 2011)[7]

power plants and drove 93 percent more vehicle miles in 2009 than 1980. US air today is the cleanest in half a century. Similar air quality improvements have been accomplished in Western Europe, Japan, Canada, Australia, and most developed nations.

Nevertheless, Climatists have redefined pollution to include carbon dioxide, now branded as "dirty carbon pollution." Since it's tough to call an invisible gas dirty, Climatists use carbon instead, hence the phrase "carbon pollution." But referring to carbon dioxide as "carbon" is as foolish as calling water "hydrogen" or naming salt "chlorine." Compounds have totally different properties than their composing elements.

Carbon itself is integral to all living things. It's the basic building block for our skin, our muscles, and our bones, and is found throughout the body of each person. Carbon accounts for about 18 percent of the human body by weight.[8] It's why humans, animals, and plants are sometimes described as "carbon-based life forms." We are full of this "dangerous carbon pollution" by natural metabolic processes.

CO$_2$ IS GREEN!

In the words of geologist Leighton Steward, "CO$_2$ is green!"[9] Carbon dioxide is plant food and fundamental to plant photosynthesis. Plants use water, CO$_2$, and sunlight to produce sugars, releasing oxygen as a waste product from the process. Carbon dioxide joins oxygen and water vapor as one of the three gases that are essential for the biosphere. Most life would not exist without carbon dioxide.

Hundreds of peer-reviewed studies show that higher levels of atmospheric CO$_2$ cause plants to grow faster and larger. Experiments show that wheat, orange trees, pine trees, hardwood trees, prairie grasses, and almost all green plants thrive in higher levels of airborne CO$_2$. Plants grow larger root systems, produce more seeds and

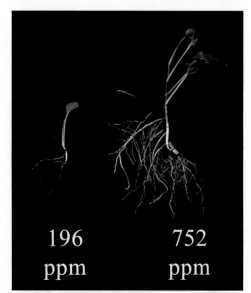

196 ppm   752 ppm

Devil's Ivy Growth and CO$_2$ Levels. Devil's Ivy or Golden Pothos plants experimentally grown at two different atmospheric CO$_2$ levels. (*CO$_2$Science*, 2012)[10]

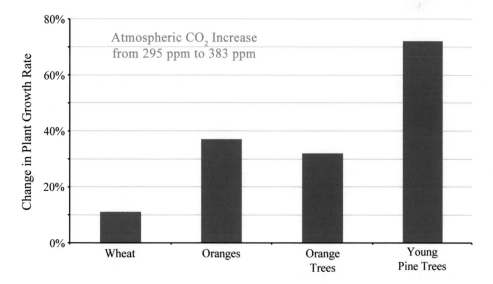

Atmospheric CO$_2$ Increase from 295 ppm to 383 ppm

Change in Plant Growth Rate

80%
60%
40%
20%
0%

Wheat   Oranges   Orange Trees   Young Pine Trees

Plant Growth with Increased CO$_2$. Calculation of improved plant growth rates from the increased atmospheric CO$_2$ level in 2007 compared to the pre-industrial CO$_2$ level. (Adapted from Robinson et al., 2007)[11]

vegetables, and bloom larger flowers with more CO_2. Tree wood density increases. Plants grow better in poor soil and in drought conditions with higher atmospheric CO_2.

Today's atmospheric CO_2 levels, just above 390 ppm, are not high by historical standards. Physicist William Happer of Princeton University reminds us:

BEWARE DANGEROUS CARBON POLLUTION![12]

About fifty million years ago…geological evidence indicates CO_2 levels were several thousand ppm, much higher than now. And life flourished abundantly.[13]

Actually, our current levels of airborne CO_2 are on the low side. At about 150 ppm, green plants become seriously starved for CO_2.

Consider that the human diet is almost entirely based on green plants or on animals that eat green plants. As Joanne Nova, Australian author, points out:

Everything on your dinner table—the meat, cheese, salad, bread, and soft drink—requires carbon dioxide to be there. For those of you who believe that carbon dioxide is a pollutant, we have a special diet: water and salt![14]

In fact, if humanity wanted to put one compound into the air that would be great for the biosphere, carbon dioxide would be that substance. Yet, almost every university and company, blinded by climate madness, tracks the size of their carbon footprint and tries to reduce their carbon emissions.

GLOBAL WARMING: A HEALTH ISSUE?

Former UK Prime Minister Tony Blair, in a speech to the US Congress in 2004, stated:

There is good evidence that last year's European heat wave was influenced by global warming. It resulted in 26,000 premature deaths and cost $13.5 billion.[15]

Mr. Blair's comments are part of another big climate whopper, that global warming is a health issue.

Emphasis on the "climate change endangers health" whopper has grown in recent years. Climate advocate groups have seized upon the health issue as a way to convince citizens who have been growing increasingly skeptical. A report by George Mason University's Center for Climate Change Communication in 2010 recommended emphasis on health:

> Successfully reframing the climate debate in the United States from one based on environmental values to one based on health values…holds great promise to help American society better understand and appreciate the risks of climate change…[16]

In addition to the usual Climatist mantra of submerging coastal cities, famine from droughts and floods, and more intense tropical storms, the health angle focuses on fears of increasing sickness and death from heat waves, air pollution, and disease. But these health fears are not only baseless, they defy common sense.

The 2009 report from the US Global Change Research Program (USGCRP) is a masterpiece of global warming propaganda. An initiative led by NOAA, it's an astonishing document of doom and gloom about disasters that will strike the United States, including droughts, floods, hurricanes, failing agriculture, water shortages, forest fires, species extinction, and even bigger and nastier poison ivy, all due to projected rising temperatures. It would take 50 pages just to refute the nonsensical claims of the report.

The report devotes a full chapter to projected US health impacts, claiming:

- increases in the risk of illness and death related to extreme heat and heat waves are very likely;

- warming is likely to make it more challenging to meet air quality standards necessary to protect public health; and

- some diseases transmitted by food, water, and insects are likely to increase.[17]

The idea that a warming planet will cause heat waves, and therefore more sickness and death, is not supported by

Poison Ivy at the Chicago Field Museum Climate Exhibit. Yes, poison ivy will flourish with more airborne CO_2. But there is no mention by the museum of the positive effects on all other plants, trees, flowers, and vegetables. (Rogers, 2010)[18]

either human experience or scientific studies. Think for a moment about your own experience. Do you suffer more illness during the cold winter or the warm summer? For most people, the answer is the winter.

According the US Center for Disease Control, the US influenza (flu) season is from November to April, during the winter months. The World Health Organization defines the Southern Hemisphere flu season to be from May to October, which are the winter months for that portion of the planet. In fact, more people get sick during cold weather than warm weather, and more people die as a result.

Scientific studies confirm our intuition on this. The late Dr. William Keating, physiology professor at Queen Mary and Westfield College, led a team that studied temperature-related deaths for people aged 65 to 74 in England, Finland, Germany, Greece, Italy, and the Netherlands. In 2000, Keating's team found that deaths related to cold temperatures were more than *nine times greater* than those related to hot temperatures in Europe. Heart attacks, strokes, and respiratory illness were responsible for most of the cold-weather deaths.[19]

Dr. Matthew Falagas, of the Alfa Institute of Medical Sciences in Athens, Greece, working with five other researchers, studied seasonal mortality for Australia, Canada, Cyprus, France, Greece, Italy, Japan, New Zealand, Spain, Sweden, and the United States. The research showed that the average number of deaths per month was lower in summer and fall months and peaked in the coldest months of the year for all nations.[20]

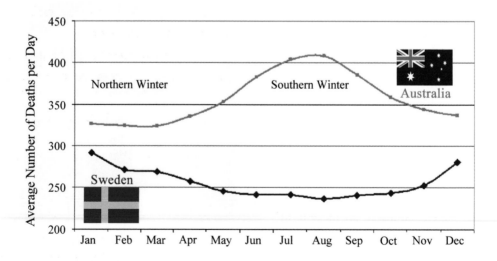

Temperature-Related Deaths in Australia and Sweden. Average number of deaths per day by month of the year for Sweden and Australia. Note the higher rate of mortality during winter months. (Falagas, 2009)[21]

Dr. Bjorn Lomborg, adjunct professor at the Copenhagen Business School, points out that any global warming that occurs will likely *reduce* human mortality:

Winter regularly takes many more lives than any heat wave: 25,000 to 50,000 each year die in Britain from excess cold. Across Europe, there are six times more cold-related deaths than heat-related deaths...by 2050...Warmer temperatures will save 1.4 million lives each year.[22]

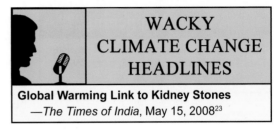

WACKY CLIMATE CHANGE HEADLINES

Global Warming Link to Kidney Stones
—*The Times of India*, May 15, 2008[23]

Despite all this evidence, the USGCRP report tried to make the case that global warming will cause more temperature-related deaths. The report states:

Some reduction in the risk of death related to extreme cold is expected...the reduction in deaths as a result of relatively milder winters attributable to global warming will be substantially less than the increase in deaths due to summertime heat extremes.[24]

Most of us have an Aunt Martha or an Uncle Jack who has retired or plans to retire to another climate. The favored locations are Alaska, Canada, and North Dakota, right? Nonsense! Senior citizens wish to retire to the warm climates of Florida, Texas, and Arizona. But don't they know that the IPCC and the US government warn about premature death in warm climates? In fact, Aunt Martha and Uncle Jack have more common sense than the climate-crazed scientists of the IPCC and the US government.

The USGCRP report tries to make the case that warmer temperatures will cause more air pollution, namely an increase in ozone, caused by "higher temperatures and associated stagnant air masses." Stagnant air masses? In earlier chapters of the same report, they warn about "more and heavier rainfall" and "more frequent and intense" extreme weather events.[25] If we believe the report, we're in for both more turbulent weather *and* stagnant air masses. As we discussed earlier, airborne ozone has been declining in the US over at least the last 30 years, despite rising temperatures. The case for higher levels of air pollution due to warmer temperatures is feeble.

"That makes climate change a bigger public health problem than AIDS, than malaria, than pandemic flu."
—US Representative Lois Capps (2011)[26]

Signs that Man-Made Warming is Happening???

Flesh-Eating Disease Is On The Rise Due To Global Warming: Experts Warn
—*Science Daily*, August 16, 2007[27]

A third issue raised in the report is the idea that disease will increase due to warmer temperatures. The report states: "Some diseases transmitted by food, water, and insects are likely to increase." It then makes a very weak case that food poisoning, West Nile Virus, parasites, and other disease-causing agents are likely to increase because of increased temperature.[28]

But most scientific studies show that temperature is a very minor factor in the spread of disease. Dr. Paul Reiter, medical entomologist at the Pasteur Institute in Paris, points out that malaria was endemic to England during the colder climate of the Little Ice Age. Reiter also recalls that the Soviet Union experienced an estimated 16 million cases of malaria during the years 1923–1925, with 30,000 cases in Archangel, a city located close to the frozen Arctic Circle.[29]

Despite rising twentieth century global temperatures of about a degree Fahrenheit, infectious diseases, including dysentery, malaria, typhoid, tuberculosis have all but been eliminated in the developed nations of the world. Improved sanitation, water purification, mosquito-abatement measures, vaccines, and public health programs have been effective in eliminating these diseases. Temperature changes are an insignificant factor.

Ignoring the lack of evidence for any real health impacts from warmer temperatures, alarm from projected changes continues to grow. A September 2010 letter to President Obama and the US Congress, signed by more than 120 leading health groups, declared that "Climate change is a serious public health issue," citing the USGCRP report.[30] Expect the climate change health whopper to be around for a long time.

POLAR BEAR PROPAGANDA

During 2011, the Philadelphia Zoo sponsored an exhibit of more than 20 endangered animals constructed out of Lego® blocks, including a life-sized sculpture of a polar bear.[31] Children learned how polar bears were endangered due to man-made global warming. Polar bear extinction is another big climate whopper that is broadcast repeatedly by the news media. The concern from global warming advocates is that arctic ice will melt, destroying the habitat of the bear, and forcing the animal to extinction.

But, there is little empirical evidence to support such concerns for the bear. According to biologist Dr. Mitch Taylor, global polar bear populations have more than doubled, from about 8,000 to 10,000 in 1960 to about 22,000 to 25,000 today.[32] If we use the logic of Climatism, CO_2 has been rising for the last 50 years and polar bear populations have been rising for the last 50 years, so increasing carbon dioxide must be producing more polar bears! But in fact, in the 1970s, the governments of Denmark, Canada, Norway, the Soviet Union, and the United States imposed polar bear hunting restrictions on the peoples of the Arctic, so the number of polar bears is growing.[33]

Climate Books[34]
We Expect to See

SAVE
POLAR BEARS

And Win
Valuable Prizes!

The IUCN Polar Bear Specialist Group issued a report in July 2009, agreeing that global bear populations total 20,000 to 25,000. But the report concluded that, of nineteen bear sub-populations in northern regions, eight were declining, three were stable, one increasing, with insufficient data on the last seven. This report showed a larger number of declining populations than their 2005 report.[35] Kassie Siegel, Director of the US Center for Biological Diversity, tells us that bears are dying as a result of climate change:

> Global warming isn't a crisis that's decades away. It's here now. The sad truth is that polar bears are already starving as global warming melts the Arctic.[36]

But detailed scientific studies do not support such alarm. A study of polar bears in the Northern Beaufort Sea, along the north coast of Alaska, was published by the US Geological Survey in 2007. The study was rigorous and extensive, using bear capture, tagging, and electronic collaring to analyze bear habits and estimate populations. Despite the fact that late summer ice declined about 10 percent per decade over the study period, researchers found that the population of bears in the area increased from about 745 in the mid-1970s to 980 in the mid-2000s, an increase of about 30 percent.[37] Since polar bears have existed for tens of thousands of years and have survived past warmer ages, it's reasonable to conclude that they can adapt to conditions of reduced arctic ice.

Comments by the Inuit people that live near the bear also differ from claims by the alarmists. Harry Flaherty, Chair of the Nunavut Wildlife Management Board, says the bear population in the Davis Strait between Canada and Greenland has more than doubled in the last ten years.[38] Gabriel Nirlungayuk, Director of Wildlife for Nunavut Tungavik, Inc.,

THE GORE EFFECT

Out With A Shiver: Global Warming Protest Frozen Out by Massive Snowfall
"Global warming activists stormed Washington Monday for what was billed as the nation's largest act of civil disobedience to fight climate change—only to see the nation's capital virtually shut down by a major winter storm."
—*FOX News*, Mar. 2, 2009[39]

points out that 40 years ago, old-timers living in the area around Hudson Bay rarely saw a polar bear. But now polar bears can be found in James Bay, at the southern tip of Hudson Bay. Nirlungayuk says:

> You now have polar bears coming into towns, getting into cabins, breaking property and just creating havoc for people up here.[40]

Despite the lack of evidence, the polar bear icon and the fable of man-made extinction is big business for some groups. In October 2011, consumer giant Coca-Cola launched its "Arctic Home" promotional campaign to promote soft drinks and to raise money to save the iconic bear. The firm introduced a white Coca-Cola® can with an image of a mother bear and her cubs. As part of the promotional campaign, the firm agreed to match up to $1 million in donations to provide a total of $2 million to the World Wildlife Fund (WWF). According to Carter Roberts, CEO of WWF:

> By working with Coca-Cola, we can raise the profile of polar bears and what they're facing, and most importantly, engage people to work with us, to help protect their home.[41]

Yes, and you can also secure big donations for the WWF. For years, Coca-Cola has donated millions to the WWF, who has profited handsomely from the polar bear whopper. Isn't it great that children all over the world are breaking open their piggy banks to send money to Coca-Cola and the WWF to save the bears?

In any case, reduction in arctic ice is being driven by natural cycles of Earth, not a trace gas in our atmosphere. Fears of polar bear decline are based only on computer model projections, supported by special interests, and not on empirical evidence.

ACID OCEANS?

In a documentary film from the Natural Resources Defense Council (NRDC), a leading environmental organization, narrator and actress Sigourney Weaver states:

> Carbon dioxide pollution is transforming the chemistry of the ocean, rapidly making the water more acidic. In decades, rising ocean acidity may challenge life on a scale that has not

occurred for tens of millions of years. So we confront an urgent choice: to move beyond fossil fuels or to risk turning the ocean into a sea of weeds.[42]

Ocean acidification is a grand whopper that is frequently delivered by climate scientists, the IPCC, and environmental groups like the NRDC. According to claims, man-made emissions of carbon dioxide are being absorbed by the oceans, where it is converted into carbonic acid, thereby changing the chemical balance of the oceans.

A solution is measured as acidic or basic (or alkaline) on a logarithmic 14-point scale, called the pH Scale. Battery acid has a pH of about one, while the base lye has a pH of as high as thirteen. Seven on the pH scale is neutral. Milk is slightly acidic, as are most of the foods that we eat.

Measured in the open ocean, sea water is basic, with a pH of about 8.2.[43] The oceans will never be acidic. What we are discussing is a fear of impacts from a reduction in alkalinity. According to computer models, doubling of atmospheric CO_2 would decrease ocean pH to about 7.9, still basic, but less so.[44] The largest concern is that such a change would destroy the coral reefs of the world. James Hansen summarizes:

> Coral reefs, the rainforest of the ocean, are home for one-third of the species in the sea. Coral reefs are under stress for several reasons, including warming of the ocean, but especially because of ocean acidification, a direct effect of added carbon dioxide. Ocean life dependent on carbonate shells and skeletons is threatened by dissolution as the ocean becomes more acid.[45]

It's true that mankind is having an impact on coral reefs and the animal populations of the seas. Overfishing and chemical pollution are important problems that need to be controlled. It's also true that carbon dioxide that is absorbed by the oceans forms carbonic acid. Nevertheless, the fear of ocean acidification from CO_2 is hugely overstated.

Dr. Glenn De'ath and others from the Australian Institute of Marine Science published a study in 2009 showing that growth rates (calcification rates) of giant coral at the Great Barrier Reef were down 14% since 1990. The study stated that the "causes remain unknown" but that the observed declines were most likely due to global warming.[46] Dr. Craig Idso, Chairman of the Center for the Study of Carbon Dioxide and Global Change, had a different interpretation of the same data. Dr. Idso pointed out that calcification rates for the giant coral were an estimated 23 percent lower in the 1500s than today. In addition, coral calcification rates increased during the last 400 years during a period of rising ocean temperatures and increasing atmospheric carbon dioxide.[47]

It's probably as difficult to develop a global average of ocean alkalinity as it is to develop

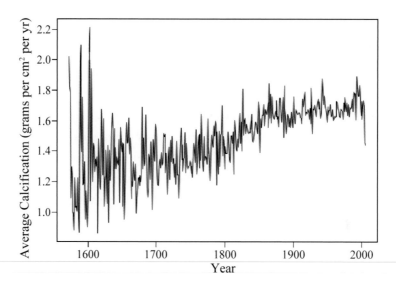

Coral Growth at Great Barrier Reef. Coral calcification rates for Porites corals on the Great Barrier Reef from 1572 to 2001. Growth rates have risen steadily over the period except for the recent downtrend after 1990. (Adapted from McIntyre, data from De'ath, 2009)[48]

a global average of surface temperature. The pH of the ocean varies by depth, becoming less basic as one goes deeper. It varies by latitude as one moves from the equator to the poles.[49] It varies by location, such as open ocean, coral reef, or kelp bed. Scientists still know little about the alkalinity of today's ocean or the oceans of past centuries.

A December 2011 study by scientists at the Scripps Institution of Oceanography found large variations in ocean pH by month, week, and even time of day. Dr. Gretchen Hoffman led a team that measured pH at fifteen locations in the Atlantic, Pacific, and Antarctic Oceans. They found that pH changes were large, from 0.1 to 1.4 units over a 30-day period. They also found that pH changed by as much as 0.35 units *over a course of days!* The study concludes that "climatology-based forecasts consistently under estimate natural variability" and that ocean residents "are already experiencing pH regimes that are not predicted until 2100" by the climate models.[50]

In fact, scuba divers will tell you that sea creatures already experience acidic pH conditions near CO_2 vents in the ocean floor. These vents bubble gas amidst coral reefs and grassy ocean pastures, as well as in millions of other locations. The fish and reefs appear to be doing quite well in these "acidified" locations, thank you!

CO_2 Vents and Healthy Coral and Sea Grass. Carbon dioxide bubbling into the ocean in the north Coral Sea from vents in Earth's crust. The coral and sea grass are healthy, despite heat from CO_2 vents that is too hot for a diver's hand. (Photographs by Halsted, 2010)[51]

97 PERCENT OF CLIMATE SCIENTISTS?

Another big whopper is the claim that the overwhelming majority of climate scientists accept that mankind is the cause of global warming. A 2009 survey by Peter Doran and Maggie Zimmerman, both of the University of Illinois, found that 97 percent of scientists who "listed climate science as their area of expertise" and who "published on the subject of climate change" answered "yes" to the question: "Do you think human activity is a significant contributing factor in changing mean global temperatures?"[52] A 2004 paper by Naomi Oreskes of the University of California found that, of 928 surveyed papers in scientific journals, "none of the papers disagreed with the consensus position" of man-made climate change.[53]

CBS News correspondent Scott Pelley hosted several reports for the network from 2004 to 2006. When asked why his reports did not acknowledge global warming skeptics, he replied:

> If I do an interview with Elie Wiesel, am I required to find a Holocaust denier?

Pelley went on to say that his team tried hard to find a respected skeptical scientist but was unable to do so.[54]

But in fact, many scientists do challenge the prevailing theory of human-caused warming. One of the earliest counters to the theory was the Heidelberg Appeal, which was released during the 1992 Rio de Janeiro Earth Summit in response to Agenda 21. Agenda 21 was the UN's "blueprint for global action" released at the Earth Summit that called for the world to adopt "sustainable development" and halt global warming. The Appeal was an opposing call to reason that science and technology are the solution to the world's problems, rather than the problem as claimed by Agenda 21. More than 4,000 scientists signed the Heidelberg Appeal.[55] Many other statements opposing the theory of man-made warming, such as the 1996 Leipzig Declaration[56] and the 2010 Climate Science Register of the International Climate Science Coalition,[57] have been largely ignored by the news media and the climate science establishment.

Probably the most significant effort was the Global Warming Petition Project, compiled by Dr. Arthur Robinson, professor of chemistry of the Oregon Institute of Science and Medicine in 2007. The Project was signed by more than 31,000 US-based scientists and over 9,000 PhDs. While not all climate experts, the scientists signed this significant petition statement:

> There is no convincing scientific evidence that human release of carbon dioxide, methane, or other greenhouse gases is causing or will, in the foreseeable future, cause catastrophic heating of the Earth's atmosphere and disruption of the Earth's climate. Moreover, there is substantial scientific evidence that increases in atmospheric carbon dioxide produce many beneficial effects upon the natural plant and animal environments of the Earth.[58]

EFFORTS TO MOBILIZE OUR YOUTH

Vladimir Ilyich Lenin, co-founder of the Soviet Union, said, "Give us the child for eight years and it will be a Bolshevik forever."[59] Today, governments are employing the techniques of past totalitarian regimes, as they feed climate change nonsense to the children of the world. Colleges, high schools, and even elementary schools now seek to train our youth to be good "eco-citizens" at an early age. Environmental science is a new favorite course for students and a fertile field for climate change propaganda, rather than the hard sciences of chemistry and physics.

The US government is heavily involved in preaching global warming nonsense. Education Secretary Arne Duncan vowed in 2010:

> Historically the Department of Education hasn't been doing enough to drive the sustainability movement, and today, I promise that we will be a committed partner in the national effort to build a more environmentally literate and responsible society.[60]

Of course, this means indoctrinating children. During a July 2011 reading event at the US Department of Education, free books featuring Nickelodeon® cartoon characters were provided to elementary school and day care-aged children. In *SpongeBob Goes Green!*, SpongeBob's friend Mr. Krabs builds a swimming pool and then SpongeBob and Mr. Krabs put carbon dioxide into the air to warm the planet and get people to "want to use my new pool all year long." In *Dora Celebrates Earth Day!*, Dora the Explorer tells her friends what they can do to "save the Earth."[61]

Big Green for Climate Change

MSU, Partners Bring Climate Change Curricula to High Schools
Part of the $2.5 million National Science Foundation Discovery Research K-12 project. Mississippi State University to create "better ways to teach climate change."
—MSU Press Release, Feb. 1, 2011[62]

Not to be outdone, NASA introduced the website *Climate Kids: NASA's Eyes on the Earth* in 2011. The site features games, "Climate Tales," and other propaganda, aimed at teaching global warming dogma to children.[63] Isn't NASA's role space exploration?

Special mention must go to *A Hot Planet Needs Cool Kids: Understanding Climate Change and What You Can Do About It*, by Julie Hall. This naked propaganda "meets the National Science Education Standards and the National Council for the Social Studies Curriculum Standards," according to the author. The book is intended to "support the development of progressive kids, so that the next generation will be prepared and motivated to care for our planet and its many forms of life." Along with a heavy dose of the usual man-made warming dogma, advice from the book includes:

A major reason people are destroying the rainforests to raise cattle is to provide hamburger meat to fast food restaurants. You and your friends and family can help by not eating fast food. It's healthier for the planet and you!

People can help reduce human overpopulation by having fewer children. Many people believe that it is best for the earth for families to have no more than one child.

Find out if your town has agreed to the Kyoto Treaty guidelines. If it hasn't, contact your mayor and urge him or her to do so.[64]

The book is intended for children as young as ten years old. Marketed to 22 middle schools in Battle Creek, Michigan, it was withdrawn after opposition from local business groups.

Second place for best exploitation of children was *Professor Schpinkee's Greenhouse Calculator* website, established by the Australian Broadcasting Corporation in 2008. The site

Act on CO$_2$. The UK government ran a TV campaign in October 2009, in which a father reads to his little girl about the horrors of CO$_2$, including the black "CO$_2$ Monster" at left.[65]

featured a kids game about the dangers of energy usage and greenhouse gases. Children were asked to take a quiz to "see how your CO$_2$ production compares to the Average Aussie greenhouse pig." Instructions then went on to say, "When you're done, click on the [skull and crossbones] to find out what age you should die at so you don't use more than your fair share of Earth's resources!"[66] The website was removed after a furor in the Australian press.

But the British government wins our prize for the best exploitation of children for its October 2009 "Act on CO$_2$" television campaign, costing British taxpayers £6 million ($9 million). The one-minute video clip shows a father reading to his frightened child in bed. As the father reads about "awful heat waves" and "terrible storms and floods," the daughter sees pictures of a black CO$_2$ monster in the sky and children and animals drowning in flooded areas. The little girl asks in fear, "Is there a happy ending?"[67] The television campaign was a desperate attempt to convince citizens that are becoming increasingly skeptical about proactive climate change policies. Climate change madness demands both the indoctrination and exploitation of children.

CLIMATE CHANGE WHOPPERS

Climate change whoppers are pervasive throughout today's society. Intellectual leaders, governments, environmental groups, universities, high schools, elementary schools, news media, and mistaken citizens around the world endlessly repeat these whoppers. Our air is being filled with dangerous carbon pollution; ozone will rise; we'll suffer premature death from heat waves; we'll be plagued with malaria, West Nile virus, and parasites; poison ivy will cover the land; the polar bears will starve; and the oceans will turn to acid. We *must* educate the children so we can avoid these disasters. Comrade Lenin would have been proud.

CHAPTER 9

BAD SCIENCE—TEMPERATURE, THE IPCC, AND REVEALING E-MAILS

"If we torture the data long enough, it will confess."
—NOBEL ECONOMICS LAUREATE RONALD COASE

Climatism is an ideology. It's an ideology with the highest of objectives—nothing short of saving Earth itself. This lofty mission has induced the world's leading climate science organizations to use slippery science to advance the theory of man-made global warming.

An in-depth look at modern climate science shows a trail of exaggeration, bias, cherry-picking, and propaganda. The very temperature records on which global warming alarm is based are fraught with error and exaggeration. The Inter-governmental Panel on Climate Change, the world's recognized climate change authority, is not objective but strongly biased toward activist climate policies.

And revealing e-mails from world's top climate scientists show a pattern of cherry pick-ing of data, deception, efforts to suppress contrary evidence, and generally poor scientific practice. Let's unfold this sorry story of science that is anything but settled.

THE SCIENCE IS SETTLED?

In Kyoto, Japan, during the 1997 Kyoto Protocol Treaty negotiations, Dr. Robert Watson, then Chairman of the IPCC, was asked about scientists who challenge United Nations con-clusions that global warming was man-made. He stated, "The science is settled…we're not going to reopen it here."[1] Thus began one of the greatest propaganda lines of Climatism.

Al Gore appeared on the CBS *Early Show* on May 31, 2006, and stated:

…the debate among the scientists is over. There is no more debate. We face a planetary emergency. There is no more scientific debate among serious people who've looked at the science…Well, I guess in some quarters, there's still a debate over whether the moon landing was staged in a movie lot in Arizona, or whether the earth is flat instead of round.[2]

Rajendra Pachauri, current Chairman of the IPCC, agrees. In an interview in 2008 with the *Chicago Tribune* he stated:

There is, even today, a Flat Earth Society that meets every year to say the Earth is flat. The science about climate change is very clear. There really is no room for doubt at this point.[3]

"Science is settled" is asserted over and over by our news media and is now a mantra of our government leaders. President Barack Obama said, "The science is beyond dispute and the facts are clear."[4] His chief of the Environmental Protection Agency, Lisa Jackson, testified in 2010 before the US Congress:

The science behind climate change is settled, and human activity is responsible for global warming.[5]

But in fact, science is never settled. Statements professing that the "science is settled" are efforts to end scientific debate and pursue political ends. Skepticism is inherent in the scientific method. Good scientists continuously test their own theories and the theories of others. Hypotheses become theories, and with enough time and testing, they can eventu-ally become laws. But the science is never settled. Newton's laws of motion were changed by Einstein's theories of relativity after more than 200 years of acceptance. Albert Einstein himself summed it up when he said:

No amount of experimentation can ever prove me right; a single experiment can prove me wrong.[6]

The phrase "science is settled" is propaganda employed to avoid open and honest discussion about the true nature of global warming. Environmental scientists at US universities use the phrase as an excuse to refuse debate on the topic. Wasn't college meant to be a forum for open exchange of ideas?

PLAYING FAST AND LOOSE WITH TEMPERATURE

Climatism is founded on the idea that the recent rise in global temperatures is abnormal. As we saw in Chapter 4, the Modern Warm Period is no different than warm periods of recent past ages. However, evidence is mounting that the temperature data sets themselves are not all they're cracked up to be. Almost all scientists agree that some cyclical warming did occur from 1975 to 1998. But it's clear that the global surface station temperature data not only contains significant errors but also major biases in favor of the theory of man-made global warming.

Developing an "average global temperature" is a questionable process. Danish physicist Bjarne Andresen pointed out that *there is no such thing* as a global temperature:

> It is impossible to talk about a single temperature for something as complicated as the climate of Earth. A temperature can be defined only for a homogeneous system. Furthermore, the climate is not governed by a single temperature. Rather, differences drive the processes and create the storms, sea currents, thunder, etc. which make up the climate.[7]

Temperature data sets compiled by the Climatic Research Unit of East Anglia University, the National Oceanic and Atmospheric Administration (NOAA), and the NASA Goddard Space Center are artificial approximations at best. They start with a patchwork collection of thousands of thermometer stations that inadequately cover the globe. Station coverage of the oceans and of the far northern and southern regions is inconsistent and poor. To cover areas without thermometers, averaging estimates are made from surrounding stations to try to fill in the holes. Political disruptions create large gaps in the temperature record. Around the year 1990, hundreds of temperature stations from the former Soviet Union stopped reporting. During the 1960s to 1980s, the network included more than 6,000 temperature stations.

But today, fewer than 1,500 stations still provide thermometer data.[8]

In addition to coverage problems, gauge measurements often contain large errors. Man-made structures such as buildings and parking lots absorb sunlight during the day and re-radiate some of this energy at night, artificially increasing local temperatures. In addition, our cars, air conditioners, and other equipment generate heat when operating. These effects are especially noticeable in cities and are called the Urban Heat Island effect. Urban warming can cause station gauges to report erroneously warm temperatures if not properly sited.

In 2007, meteorologist Anthony Watts set out to measure the effect of paint changes on thermometer stations that are used by NOAA to track US temperatures. In the process, he found several stations that did not meet the guidelines established by the National Climatic Data Center (NCDC), a NOAA organization. According to NCDC guidelines, excellent sites are to have "grass/low vegetation ground cover" and be located "at least 100 meters from an artificial heating or reflecting surface." Poor sites are those with a "temperature sensor located next to/above an artificial heating surface such as a building, roof top, parking lot, or concrete surface."[9]

Since the NCDC did not have a temperature site audit program, Watts and Dr. Roger Pielke Sr. of the University of Colorado initiated a larger project. They established a volunteer effort to photographically audit US temperature measuring stations. More than

Poorly Sited Temperature Station. The photograph at left shows a temperature station in Perry, Oklahoma, that is located next to a fire station and parking lot. The infrared image at right shows that heat from the building is biasing the measurements of the station. Watts (2009)[10]

650 volunteers participated in the effort, and 1,007 of 1,221 stations were audited by the summer of 2011. The results of the survey were striking.

The study found that more than 700 US temperature stations, over 70 percent of the sites rated, were found to be "poor" to "very poor" according to the NCDC's own rating system. These stations are likely to have temperature errors as large as 2°C (3.6°F).[11] Trend analysis by Watts and others concluded that many of these errors cancel out, but it's clear that the US surface temperature record is unreliable. Yet, the US data set is regarded as the best in the world, with data from other nations likely to be worse.

The summer of 2011 was a hot one for the central US. Many locations sweltered under "triple-digit high temperatures" (over 100°F) for weeks at a time. Climatists chuckled, asking, "Where are the climate skeptics now?" But a look at history shows that these temperatures were far below most state high-temperature records. Data from NOAA shows that only two state high-temperature records were set in the last 18 years. In contrast, 23 state highs

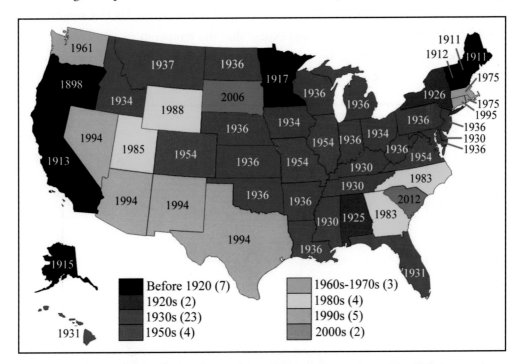

Year and Decade of US State High-Temperature Records. Year of the high-temperature record is shown for each state. Colors show records by decade or time period. Only two state high-temperature records were set during the last 18 years, while 23 records were set during the decade of the 1930s. More than two-thirds of the hot records were set before 1960. No records were set during the 1940s. (NOAA, 2013)[12]

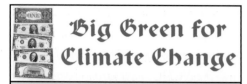

Big Green for Climate Change

The National Oceanic and Atmospheric Adminstration, a division of the US Department of Commerce, spent an estimated $378 million for climate change in Fiscal Year 2010.[13]

were set in the 1930s, and more than two-thirds of the state high-temperature records were set before 1960.[14] In contrast, NOAA and NASA temperature data sets show steadily rising US temperatures throughout the last century. How can it be that state high-temperature thermometer records differ so markedly from US government temperature data sets?

The answer is that the US government makes "adjustments" to the raw temperature data. According to a 2008 NOAA paper, after the raw thermometer data is received, a computer algorithm is used to "homogenize" the data, adjusting for time-of-observation, station moves, thermometer types, and other factors, to arrive at an "accurate, unbiased, and modern historical climate record," according to NOAA.[15] This sounds good, until you look at the adjustment, which is shown in the figure below. If the NOAA modifications were accurate and unbiased, one should expect that the temperature corrections between stations to roughly balance out. Instead, NOAA adds little adjustment to data from 1900

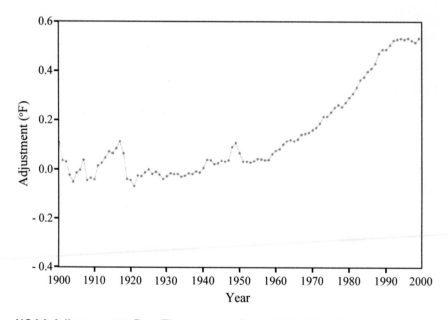

NOAA Adjustment to Raw Thermometer Data 1900–1999. Raw thermometer measurements have been adjusted upward by the US government, particularly since 1960. (Williams, et. al., 2007)[16]

to 1960, but after 1960, they add an upward adjustment that rises to 0.5°F (0.3°C) by the year 2000. *So NOAA is adding a half-degree upward temperature bias to the raw thermometer data!* This gives a whole new meaning to the phrase "man-made global warming."

The United States is not the only government injecting warming bias into temperature data sets. Ken Stewart, a climate researcher in Australia, analyzed the data from the 100 recording stations of the Australian National Climate Centre, Bureau of Meteorology. He found that the raw thermometer data showed a rise of 0.6°C per century, while the "corrected" data showed a greater rise of 0.85°C per century. A look at the adjustment shows that the thermometer readings from 1910 to 1950 have been adjusted downward, boosting the upward trend in temperatures.[17]

In nearby New Zealand, The New Zealand Climate Science Coalition (NZCSC) conducted a similar review of the homogenization of that nation's temperature data, officially held by the National Institute of Water and Atmospheric Research (NIWA). They found that NIWA made major adjustments to data from seven of New Zealand's official weather stations. Data from six stations were adjusted upward and only one was reduced. Why are these adjustments *almost always upward* in support of the theory of man-made global warming?

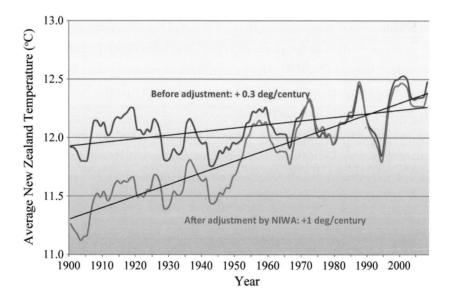

Raw and Adjusted Temperature Record for New Zealand 1900–2008. After adjustment by the National Institute of Water and Atmospheric Research (NIWA), the adjusted temperature record (red) for New Zealand shows a larger temperature rise than the raw thermometer data (green). (Leyland, 2010)[18]

The Coalition found that the raw thermometer data showed a small rise of 0.3°C per century, but after adjustment, the temperature trend became a whopping 1.0°C per century![19] The temperature adjustments were "prepared in consultation with" Jim Salinger, a former NIWA employee, and a lead author for the Intergovernmental Panel on Climate Change. In a press release, the NZCSC remarked:

Climategate Capers

"…We have 25 or so years invested in the work. Why should I make the data available to you, when your aim is to try to find something wrong with it…"
—Phil Jones, Director of the Climate Research Unit of East Anglia University, responding to a request for temperature station data by Australian climate researcher Warwick Hughes (2004)[20]

> We have discovered that the warming in New Zealand over the past 156 years was indeed man-made, but it had nothing to do with emissions of CO_2—it was created by man-made adjustments of the temperature.[21]

The Coalition brought suit in the New Zealand High Court, asking the court to invalidate the official temperature record of NIWA. The incident became known as "Kiwigate" in the New Zealand press.[22]

In a surprising move, the New Zealand government filed a Statement of Defense in the High Court, disavowing the country's official climate record. Despite the fact that NIWA was a government-owned organization, that NIWA had been advising the NZ government on climate issues, and that the temperature record had been officially called the New Zealand Temperature Record (NZTR), the government announced that New Zealand "did not have an official temperature record." They further stated that the NZTR was unofficial and strictly for internal research purposes.[23]

The NOAA dataset, the Australian temperature record, and the NZTR are provided to the Climate Research Unit (CRU) at UK East Anglia University to form part of the HadCRUT3, the leading global surface temperature dataset. The CRU then makes further "adjustments" to the temperature data it receives from nations around the world. And some of these adjustments are also questionable.

In December 2009, shortly after the breaking of the Climategate scandal (see page 158), the Moscow Institute for Economic Analysis (IEA) issued a paper titled, "How Warming is Being Made: The Case of Russia." The paper claimed that the CRU had "cherry-picked" Russian climate data. The IEA pointed out that the CRU had used data from only 25 percent of Russian temperature stations and that over 40 percent of the nation's territory was not included, even though there was no lack of meteorological stations and observations from these areas. The IEA charged that the CRU database included some stations with incomplete data that highlighted the warming process, rather than stations with more complete data that did not show warming.[24]

It's clear that global surface temperature records are unreliable at best and deliberately biased at worst. Today, data sets are based on only one-quarter of the temperature stations that were reporting in 1980. Significant coverage gaps exist over oceans and colder global regions. Two-thirds of US temperature stations fail National Climatic Data Center site guidelines and suffer from significant urban heat island bias. Man-made adjustments to temperature data from Australia, New Zealand, Russia, and the United States have been shown to boost warming trends in the temperature record. As summarized by meteorologists Anthony Watts and Joseph D'Aleo:

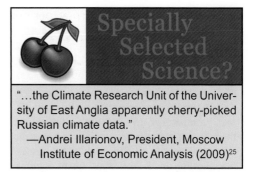

Specially Selected Science?

"...the Climate Research Unit of the University of East Anglia apparently cherry-picked Russian climate data."
—Andrei Illarionov, President, Moscow Institute of Economic Analysis (2009)[25]

> These factors all lead to significant uncertainty and a tendency for overestimation of century-scale temperature trends. A conclusion from all findings suggest that global data bases are seriously flawed and can no longer be trusted to assess climate trends or rankings or validate model forecasts. And consequently, such surface data should be ignored for decision making.[26]

BIAS AT THE IPCC

The Intergovernmental Panel on Climate Change of the United Nations is regarded as the accepted scientific authority on climate change by a misguided mankind. The IPCC's stated mission is:

> ...to provide the world with a clear scientific view on the current state of knowledge in climate change and its potential environmental and socio-economic impacts.[27]

The IPCC claims to be a scientific body:

> The IPCC is a scientific body. It reviews and assesses the most recent scientific, technical and socio-economic information produced worldwide relevant to the understanding of climate change. It does not conduct any research nor does it monitor climate related data or parameters.[28]

It also claims to be "balanced" and "policy-neutral":

YOU CAN'T HANDLE THE TRUTH!

"On one hand, as scientists we are ethically bound to the scientific method, in effect promising to tell the truth, the whole truth, and nothing but—which means that we must include all the doubts, the caveats, the ifs, ands, and buts. On the other hand, we are not just scientists but human beings as well. And like most people we'd like to see the world a better place, which in this context translates into our working to reduce the risk of potentially disastrous climatic change. To do that we need to get some broad-based support, to capture the public's imagination. That, of course, entails getting loads of media coverage. So we have to offer up scary scenarios, make simplified, dramatic statements, and make little mention of any doubts we might have. This 'double ethical bind' we frequently find ourselves in cannot be solved by any formula. Each of us has to decide what the right balance is between being effective and being honest."

—Dr. Stephen Schneider, former IPCC Coordinating Lead Author (1989)[29]

Because of its scientific and intergovernmental nature, the IPCC embodies a unique opportunity to provide rigorous and balanced scientific information to decision makers. By endorsing the IPCC reports, governments acknowledge the authority of their scientific content. The work of the organization is therefore policy-relevant and yet policy-neutral, never policy-prescriptive.[30]

However, these claims are just not true. The IPCC is *not* a scientific organization. Instead, it is a political organization with the unstated mission to convince humanity that greenhouse gases are destroying Earth's climate.

From the start, *a balanced look at climate change was never a serious consideration by the IPCC.* The organization never objectively analyzed whether climate change was primarily driven by natural forces, or whether global warming might even be a net positive for humanity. The IPCC held its first plenary meeting in November 1988. That very same month, the UN General Assembly adopted a resolution charging the IPCC to provide a "comprehensive review and recommendations" on climate change by 1990. Under the gun, the IPCC immediately formed three working groups: WGI (Science), WGII (Impacts), and WGIII (Response Strategies) in an effort to complete their First Assessment Report by 1990. John Zillman, principal delegate from Australia to the IPCC, revealed that it was necessary for each of the working groups to "work in parallel rather than in sequence."[32] For the Impact and Response Strategies working groups to write their reports, *the conclusion that man-made emissions were the cause of global warming must have already been accepted.* Balance was never a consideration.

Big Green for Climate Change

President Obama requested $13 million for the Intergovernmental Panel on Climate Change in the FY2012 budget.[31]

Second, the IPCC claims to be a scientific organization but then pushes for "consensus" in support of a political position. This is done by selecting lead authors who support the theory of man-made global warming, and then giving them authoritarian editorial power. The lead authors then control the content of the final report. Dr. Richard Lindzen of MIT observes that the IPCC process is not like a standard scientific peer-reviewed process:

> Under true peer-review…a panel of reviewers must accept a study before it can be published in a scientific journal. If the reviewers have objections the author must answer them or change the article to take reviewers' objections into account. Under the IPCC review process, the authors are at liberty to ignore criticisms.[33]

Dr. Fredrick Seitz, formerly one of America's most distinguished scientists and physicists, specifically criticized the process to produce the Summary for Policy Makers section of the 1995 Second Assessment Report. Dr. Seitz charged that lead authors made key changes to the report after scientists had accepted the text:

> But more than 15 sections in Chapter 8 of the report—the key chapter setting out the scientific evidence for and against a human influence over climate—were changed or deleted after the scientist charged with examining this question had accepted the supposedly final text…The following passages are examples of those included in the approved report but deleted from the supposedly peer-reviewed published versions:

- "None of the studies cited above has shown clear evidence that we can attribute the observed [climate] changes to the specific cause of increases in greenhouse gases."

- "No study to date has positively attributed all or part [of the climate change observed to date] to anthropogenic [man-made] causes."

- "Any claims of positive detection of significant climate change are likely to remain controversial until uncertainties in the total natural variability of the climate systems are reduced."

Dr. Seitz summarized:

> In my more than 60 years as a member of the American scientific community, including service as president of both the National Academy of Sciences and the American Physical Society, I have never witnessed a more disturbing corruption of the peer-review process than the events that led to this IPCC report.[34]

Despite these criticisms, the IPCC continues to claim that their reports are based on peer-reviewed science. Dr. Rajendra Pachauri, Chairman of the IPCC, has repeatedly made this claim. Pachauri stated in a 2007 interview:

> The IPCC doesn't do any research itself. We only develop our assessments on the basis of peer-reviewed literature.[35]

But recent developments have shown this to be far from the truth.

Chapter 10 on Asia, of the Nobel Prize-winning 2007 Fourth Assessment Report (AR4), contained this alarming statement about Himalayan glaciers:

> Glaciers in the Himalaya are receding faster than in any other part of the world...the likelihood of them disappearing by the year 2035 and perhaps sooner is very high if the Earth keeps warming at the current rate.[36]

Al Gore used this statement to forecast that 40 percent of the world's population would face a drinking water shortage within the next 50 years. UK Prime Minister Gordon Brown reinforced this fear at the 2009 Copenhagen Climate Conference:

> ... in just 25 years the glaciers in the Himalayas which provide water for three-quarters of a billion people could disappear entirely.[37]

But also in 2009, Professor Graham Cogley, a glacier expert at Trent University in Canada, questioned the Himalaya glacier melt forecast, estimating that the 2035 date was wrong by more than 300 years.[38] India's environment minister, Jairam Ramesh, then released a study that determined that Himilayan glaciers may not be melting as much due to global warming as was feared. He accused the IPCC of being "alarmist." IPCC Chairman Pachauri attacked the report, labeling it "voodoo science."[39]

But soon afterward, Dr. Murari Lal, the coordinating lead author for Chapter 10, admitted that he was aware that the "melting by 2035" statement was not based on peer-reviewed literature. He further admitted that *the statement was included to try to put political pressure on world leaders*:

> We thought that if we can highlight it, it will impact policy-makers and politicians and encourage them to take some concrete action.[40]

Dr. Pachauri subsequently apologized for the "melting by 2035" claim.

Chapter 13 of the AR4 discussed Latin America and predicted catastrophic changes for the Amazon rainforest:

Up to 40% of the Amazonian forests could react drastically to even a slight reduction in precipitation…forests will be replaced by ecosystems that have more resistance to multiple stresses caused by temperature increase, droughts and fires, such as tropical savannas.[41]

This claim that 40 percent of Amazonian forests would disappear due to climate change was again not from a peer-reviewed source. The IPCC cited a paper by two environmental organizations, the World Wildlife Fund (WWF) and the International Union for Conservation of Nature. But the WWF later denied the claim that 40 percent of Amazon forests would disappear. The basis for the 40 percent text was finally traced to a website of a Brazilian environmental advocacy group.[42]

In yet another example, the Synthesis Report of the 2007 AR4 paints a dire picture for Africa due to climate change:

By 2020, in some countries, yields from rain-fed agriculture could be reduced by up to 50%. Agricultural production, including access to food, in many African countries is projected to be severely compromised. This would further adversely affect food security and exacerbate malnutrition.[43]

The IPCC cites the source of the claim as "Agoumi, 2003." The author, Ali Agoumi, is not a climate scientist, and his report contains no primary research. It is a non-peer-reviewed summary report about three countries: Algeria, Morocco, and Tunisia. The IPCC extended this non-peer-reviewed report about three nations to warn of looming catastrophe for the whole continent of Africa.[44]

After the IPCC Himalayan glacier, Amazon forest, and African agriculture claims were exposed to be unfounded, Canadian author Donna Laframboise organized a group to conduct a more thorough analysis of IPCC AR4 sources. A team of 43 volunteers from twelve countries examined the thousands of IPCC references that were listed at the end of each chapter. Of the 18,531 references cited in the 2007 IPCC report, 5,587, or 30 percent, were found to be from the "gray literature," or non-peer-reviewed. Twenty-one of the 44 chapters of the AR4 relied on peer-reviewed sources less than 60 percent of the time.[45]

To anyone who has studied the history of the IPCC, it's clear that the assessment reports are heavily biased. Selective choice of science, use

FAILED PREDICTIONS

"Fifty Million Climate Refugees by 2010" —United Nations Environment Programme website, 2005 (now erased from site)[46]

of non-peer-reviewed sources, and rewriting of report chapters by lead authors was used to build a story that man-made emissions are causing catastrophic climate change. With such techniques, the IPCC has been able to persuade mankind of the coming climate disaster.

Some have proposed that the IPCC be reorganized to remove the bias and restore the credibility of the organization. Dr. Roy Spencer warns against this:

> Unfortunately, there is no way to "fix" the IPCC, and there never was. The reason is that its formation over 20 years ago was to support political and energy policy goals, not to search for scientific truth. I know this not only because one of the first IPCC directors told me so, but also because it is the way the IPCC leadership behaves. If you disagree with their interpretation of climate change, you are left out of the IPCC process. They ignore or fight against any evidence which does not support their policy-driven mission, even to the point of pressuring scientific journals not to publish papers which might hurt the IPCC's efforts.[47]

THE CLIMATEGATE SCANDAL

On November 19, 2009, an unidentified hacker or internal whistle-blower downloaded more than 1,000 documents and e-mails from the Climatic Research Unit (CRU) at East Anglia University and posted them on a server in Russia. Within hours, these documents were accessed by websites around the world. The e-mails were a subset of confidential communications between top climate scientists in the UK, the US, and other nations over the last fifteen years. These were the very same scientists that developed the surface temperature data sets, promoted the Mann Hockey Stick Curve, and wrote and edited the core of the IPCC assessment reports. The incident was branded "Climategate" by British columnist James Delingpole, a label soon adopted by the world. These e-mails provide us with insight into practices by researchers that are poor science at best and fraudulent at worst. The e-mails were released on the eve of the 2009 United Nations Climate Conference in Copenhagen.

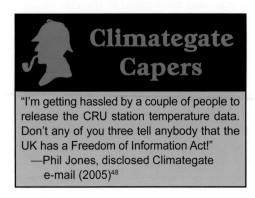

Climategate Capers

"I'm getting hassled by a couple of people to release the CRU station temperature data. Don't any of you three tell anybody that the UK has a Freedom of Information Act!"
—Phil Jones, disclosed Climategate e-mail (2005)[48]

Let's go back to the fall of 1999. Lead authors for the IPCC are engaged in an e-mail discussion on how to portray important "climate reconstruction data" in the upcoming

Third Assessment Report (TAR), scheduled for publication in 2001. Key players include Dr. Phil Jones, Director of the Climatic Research Unit of the University of East Anglia, Dr. Keith Briffa, also of the CRU, Dr. Chris Folland of the UK Meteorological Office, Dr. Michael Mann from the University of Virginia, and other leading climate scientists.

Mann, Jones, and Briffa had recently completed analyses of tree ring data and other temperature proxies in an effort to try to reconstruct estimates of Earth's surface temperatures for the past 1,000 years. Recall that tree rings can be a proxy, an indicator of past temperature trends. Climate scientists in the US and UK had been hard at work constructing curves of past temperatures in an effort to demonstrate that modern temperatures were abnormally warm. As we showed in Chapter 4, these new analyses ignored dozens of other studies that showed the natural climatic periods of the Medieval Warm Period and the Little Ice Age.

"...it would be nice to try to 'contain' the putative 'MWP' [Medieval Warm Period]..."
—Michael Mann, IPCC lead author, disclosed Climategate e-mail (2003)[49]

The officers of the IPCC are excited about this work. In a September 22, 1999 e-mail, Chris Folland said:

> A proxy diagram of temperature change is a clear favourite for the Policy Makers Summary.[50]

It's clear that the IPCC wants to present a powerful story that the twentieth century warming was unusual. Keith Briffa mentions this in an e-mail that same day:

> I know there is pressure to present a nice tidy story as regards "apparent unprecedented warming in a thousand years or more in the proxy data" but in reality the situation is not quite so simple.[51]

Indeed, there is a problem with the data. While the temperature reconstructions developed by Mann and Jones showed a temperature upturn near the end of the twentieth century, the Briffa reconstruction showed a distinctive downturn (however, remember from Chapter 4 that Mann's algorithm was inherently biased to produce a hockey-stick rise). The Briffa series posed a significant problem, as stated in Chris Folland's September 22 e-mail:

> But the current diagram with the tree ring only data [the Briffa reconstruction] somewhat contradicts the multi-proxy data and dilutes the message rather significantly.[52]

Michael Mann agrees in another e-mail the same day:

Keith's [Briffa] series…differs in large part in exactly the opposite direction that Phil's [Jones] does from ours. This is the problem we all picked up on (everyone in the room at IPCC was in agreement that this was a problem and a potential distraction/detraction from the reasonably consensus viewpoint we'd like to show w/ the Jones et al and Mann et al series).[53]

Here we see evidence that the lead authors of the IPCC are trying to promote the "story" of abnormal modern warming. After much e-mail communication over the next two months between Jones, Briffa, Mann, Folland, Tom Karl of NOAA, and others, this "hockey-stick team" arrived at a solution.

They fudged the data. First, the proxy curves themselves did not show a big temperature increase at the end of the twentieth century. So, to create a more powerful appearance, (the sharp upward blade of the hockey stick), they pasted the "instrumental data" of the modern temperature record onto the end of the proxy data. This is captured in the most famous of the Climategate e-mails, penned by Phil Jones on November 16, 1999, and sent to Mann, Briffa, and others:

> I've just completed Mike's [Mann] Nature trick of adding in the real temps to each series for the last 20 years (i.e. from 1981 onwards) and from 1961 for Keith's [Briffa] *to hide the decline.* [emphasis added][54]

Second, the hockey-stick team solved the "Briffa problem" by deleting the data after 1960 that showed a decline in the Briffa tree-ring proxy data series. Rather than admit that the decline in the series for the last of the twentieth century (when temperatures were rising) made the whole tree-ring analysis questionable, the hockey-stick team decided to "hide the decline." Figure 2.21 of the IPCC Third Assessment Report (TAR) shows the truncated Briffa series and references the Briffa (2000) paper, but the TAR *does not mention* that they shortened the Briffa curve. Data from disclosed climategate e-mails shows the full Briffa series, including a downturn in the curve after 1960, opposite the rise in global temperatures.

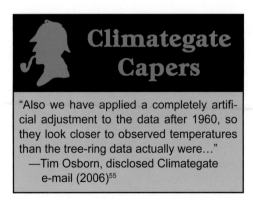

Climategate Capers

"Also we have applied a completely artificial adjustment to the data after 1960, so they look closer to observed temperatures than the tree-ring data actually were…"
—Tim Osborn, disclosed Climategate e-mail (2006)[55]

In addition to the questionable modification of data that created the hockey stick, the e-mails displayed numerous examples of bias toward the theory of man-made warming by scientists at the CRU, NASA, NOAA,

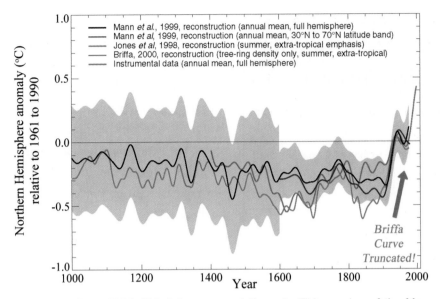

Figure 2.21 from IPCC Third Assessment Report. This version of the Mann Hockey Stick Curve shows four proxy series completed by the twentieth century temperature record (instrumental data). The IPCC does not tell you that the Briffa data (green line) has been cut off after about 1960 ("Briffa Curve Truncated" note added). (Adapted from IPCC, 2001)[56]

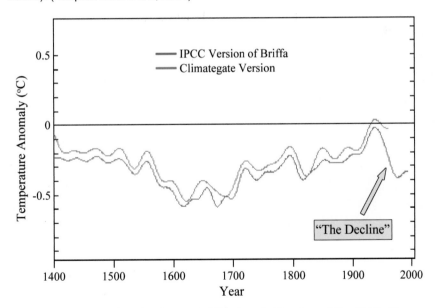

Comparison of IPCC Version and Climategate Version of the Briffa Curve. The green curve is the IPCC curve used in the 2001 TAR with data cut off after 1960. The red curve is a reconstruction of the full Briffa curve from data found in released Climategate e-mails. (Adapted from McIntyre, 2009)[57]

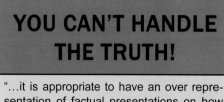

YOU CAN'T HANDLE THE TRUTH!

"...it is appropriate to have an over representation of factual presentations on how dangerous it is, as a predicate for opening up the audience to listen to what the solutions are, and how hopeful it is that we are going to solve this crisis."
—Al Gore (2006)[58]

the University Corporation for Atmospheric Research (UCAR), and other major scientific organizations and universities. The e-mails were seriously at odds with accepted scientific procedure. Evidence of data manipulation, evasion of freedom of information (FOI) requests, deliberate deletion of e-mails to avoid sharing of information, and efforts to bias the peer-review literature process were apparent in the disclosures.

Senders or receivers of the Climategate e-mails included:

- Phil Jones, Director of the CRU, and Peter Stott, Keith Briffa, Tim Wrigley, Tom Osborn, and Chris Folland of the Hadley Centre CRU
- James Hansen and Gavin Schmidt of NASA
- Thomas Karl, Director of NOAA, and Susan Solomon and Thomas Peterson of NOAA
- Kevin Trenberth and Tom Wigley of the University Corporation for Atmospheric Research
- Benjamin Santer of Lawrence Livermore Laboratory
- Jonathan Overpeck and Malcolm Hughes of the University of Arizona
- Myles Allen of the University of Oxford
- Michael Mann of the University of Virginia

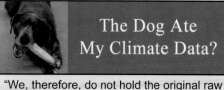

The Dog Ate My Climate Data?

"We, therefore, do not hold the original raw data but only the value-added (i.e. quality controlled and homogenized) data."
—Climatic Research Unit, the world's leading provider of global temperature data, admitting that it can't produce the original thermometer data.[59]

Each of these scientists played a major role as contributing authors, lead authors, or editors of the IPCC reports. This team that wrote the IPCC reports was anything but objective.

To top it all off, after multiple and repeated requests for raw temperature data by scientists around the world, the CRU admitted in 2009 that it does not have the original temperature data for the HadCRUT3 global temperature data set. They claim to have only kept the homogenized data that has been adjusted by CRU scientists. In effect, they are saying "trust us" about the twentieth century warming, for which the world is spending billions to "fight man-made climate change."

THE INQUIRY WHITEWASH

But an additional astonishment was yet to come. After the press, the universities, the scientific organizations of the world, and Her Majesty's Government read the Climategate e-mails, calls rang out for an investigation of the science at the Climatic Research Unit. Three formal inquiries were established in the UK to get to the bottom of the Climategate e-mails.

The House of Commons Science and Technology Committee sent a letter to the University of East Anglia (UEA) on December 1, 2009, requesting an explanation of the Climategate affair. UEA indicated that it would set up its own "independent inquiry." The House Science and Technology committee also initiated its own investigation on January 22, 2010.[60]

After just five weeks of cursory review the House committee published a report on March 31, 2010, finding that:

> In the context of the sharing of data and methodologies, we consider that Professor Jones' actions were in line with common practice in the climate science community…We are content that the phrases such as "trick" or "hiding the decline" were colloquial terms used in private e-mails and the balance of evidence that we have seen does not suggest that Professor Jones was trying to subvert the peer review process…a breach of the Freedom of Information Act 2000 may have occurred but that a prosecution was time barred…[61]

The House of Commons did not really look deeply into the whole Climategate affair and apparently wanted to leave the work to the

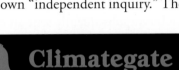

Climategate Capers

"…I can't see either of these papers being in the next IPCC report. Kevin [Trenberth] and I will keep them out somehow, even if we have to redefine what the peer-review literature is!"
—Phil Jones, IPCC lead author, disclosed Climategate e-mail (2004)[62]

See No, Hear No, Speak No Climate Evil!

"…the evidence we have seen does not suggest that Professor Jones was trying to subvert the peer review process."
—House of Commons Science and Technology Committee(2010)[63]

two other inquiries that were being initiated by the University of East Anglia (UEA).

On March 22, the UEA announced a Scientific Appraisal Panel to be chaired by Lord Ronald Oxburgh, consisting of six academics from universities. The UEA announced that the panel would perform an "independent reappraisal of the science."[64] After a deliberation of only three weeks, the panel issued a tiny five-page report, finding "no deliberate scientific malpractice" but criticizing the CRU for being "slightly disorganized."[65] Amazingly, the panel only interviewed members of the CRU and only reviewed technical papers published by members of the CRU. Independent skeptical scientists such as Stephen McIntyre, who understood the manipulation that appeared to have occurred, were not interviewed.

During the Oxburgh Inquiry interview of Phil Jones, Jones reportedly stated that:

> …it was probably impossible to do these [1000-year temperature] reconstructions with any accuracy.

McIntyre subsequently sent an e-mail to Lord Oxburgh, requesting an addendum to the report to add this comment by Jones, which was not in the issued panel report. Lord Oxburgh replied, "…the science was not the subject of our study."[66] Hmmm. So, the science was not the subject of the Scientific Appraisal Panel?

The Oxburgh Inquiry was neither scientific nor independent, as the university had promised. Lord Oxburgh was CEO of the Carbon Capture and Storage Association and Chairman of Falck Renewable Resources, organizations with strong vested interests in activist climate policy. Two of the panel members, Kerry Emanuel and Lisa Graumlich, were co-authors of technical papers with Michael Mann and Malcolm Hughes, Climategate e-mail authors. The technical papers to be reviewed by the panel were vetted by Phil Jones, himself. The Oxburgh Inquiry was hardly independent in either membership or material reviewed.[68]

The scanty findings of the Oxburgh Inquiry were roundly criticized. Climate scientist Judith Curry remarked:

> When I first read the report, I thought I was reading the executive summary and proceeded to look for details; well there weren't any. And I was concerned that the report explicitly did not address the key issues that had been raised by the skeptics.[69]

Climategate Capers

"Mike, can you delete any e-mails you may have had with Keith re AR4? Keith will do likewise…Can you also e-mail Gene and get him to do the same? I don't have his e-mail address…We will be getting Caspar to do likewise."
—Phil Jones, disclosed Climategate e-mail (2008)[67]

Member of Parliament Graham Stringer stated, "The Oxburgh Report looks much more like a whitewash."[70]

The third panel, The Independent Climate Change E-mails Review, was established by UEA under Sir Muir Russell to examine the released e-mails, the policies and practices of the Climatic Research Unit, and Freedom of Information Act Issues. This was the second panel set up by the university to investigate itself. Like the Oxburgh Inquiry, the Russell Inquiry would not assess the climate science of the CRU.

Also, like the Oxburgh Inquiry, the Russell Inquiry was tainted by lack of independence issues. One of the five panelists, Philip Campbell, editor of the pro-manmade-warming journal *Nature*, resigned after it was revealed that he had given a radio interview after the Climategate scandal broke in which he judged the scientists blameless. Another panelist, Geoffery Boulton, was a former employee of the University of East Anglia Department of Environmental Sciences. Boulton also signed a UK Meteorological Office petition in December of 2009, supporting the integrity of the CRU scientists. Boulton did not resign, despite the fact that the panel was established in 2010 to assess the integrity of those same scientists.[71]

Like the Oxburgh Inquiry, the Russell Inquiry did not hold any public hearings. The panel interviewed CRU and UEA staff but did not interview any climate skeptics. A 160-page report was issued on July 7. The Russell Inquiry exonerated the scientists of the Climatic Research unit, stating:

> ...we find that their rigor and honesty as scientists are not in doubt...we did not find any behaviour that might undermine the conclusions of the IPCC assessments...But we do find that there has been a consistent pattern of failing to display the proper degree of openness...[73]

The Russell Inquiry was a remarkable exercise in either bias, ineptitude, or both.

Climategate Capers

"...If you look at the attached plot you will see that the land also shows the 1940s warming blip (as I'm sure you know). So, if we could reduce the ocean blip by, say 0.15 deg C, then this would be significant for the global mean—but we'd still have to explain the land blip..."
—Tom Wigley, University Corporation for Atmospheric Research, disclosed Climategate e-mail to Phil Jones (2009)[72]

See No, Hear No, Speak No Climate Evil!

"On the allegation of withholding temperature data, we find that CRU was not in a position to withhold access to such data or tamper with it."
—Muir Russell Review (2010)[74]

"Selective inquiry" was the course of action. For example, on the issue of reconstruction of past temperatures, the panel focused on the IPCC Fourth Assessment Report. This approach completely ignored the manipulation of the Mann Hockey Stick and the shortening of the Briffa Curve for the IPCC Third Assessment Report. Most astonishing, Russell and his team *never even asked* Jones and his colleagues whether they had actually deleted e-mails, let alone why they had done so.[75]

In sum, the three Climategate inquiries did a remarkably poor job. No skeptical scientists were interviewed. No transcripts of inquiry proceedings were made. They discounted written e-mail evidence available for all to see. Ross McKitrick of the University of Guelph in Canada, who provided papers to the Muir Russell Review, summarizes what is still needed regarding a proper Climategate investigation:

> The world still awaits a proper inquiry into climategate: one that is not stacked with global warming advocates and one that is prepared to cross-examine evidence, interview critics as well as supporters of the CRU and other IPCC players, and follow the evidence where it clearly leads.[76]

ANOTHER CLIMATEGATE E-MAIL RELEASE

On November 22, 2011, two years after the initial Climategate e-mail release, a second batch of 5,200 e-mails was delivered to skeptical websites by the unknown hacker or whistle-blower. This second e-mail release covered the same time period as the first release, and was confirmed to be genuine by the University of East Anglia. The second release appeared to be timed to cause maximum impact prior to the UN Climate Conference in Durban, South Africa, beginning on November 28.

The release was accompanied by a message from those responsible, discussing their motivations for the document delivery:

> Over 2.5 billion people live on less than $2 a day. Every day nearly 16,000 children die from hunger or related causes. One dollar can save a life—the opposite must also be true. Poverty is a death sentence. Nations must invest $37 trillion in energy technologies by 2030 to stabilize their greenhouse gas emissions at stainable levels. Today's decisions must be made on all the information we can get, not on hiding the decline.[77]

Like the first batch of released e-mails, the new additional release shows a pattern of bias, efforts to avoid freedom of information requests, a "bunker mentality" against skeptical scientists, as well as efforts to control the peer-review process for climate science publications. But the release provides additional insight into the thinking and the efforts of activist climate scientists to persuade the world about catastrophic global warming. Several leading scientists in the Climategate e-mail net referred to efforts to convince the world of man-made global warming as "the cause." These include Dr. Joseph Alcamo, Director of the Center for Environmental Systems Research at the University of Kassel, Germany, Ian Harris of the Climatic Research Unit, and Michael Mann. In an e-mail to Phil Jones in 2004, Mann states:

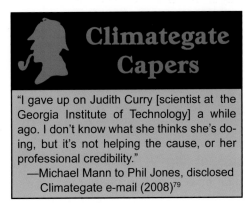

Climategate Capers

"I gave up on Judith Curry [scientist at the Georgia Institute of Technology] a while ago. I don't know what she thinks she's doing, but it's not helping the cause, or her professional credibility."
—Michael Mann to Phil Jones, disclosed Climategate e-mail (2008)[79]

> By the way, when is Tom C [Tom Crowley of Texas A & M University] going to formally publish his roughly 1500 year reconstruction??? It would help *the cause* to be able to refer to that reconstruction as confirming Mann and Jones, etc. [emphasis added][78]

Phil Jones, head of the CRU, is found to have authored further communications that propose hiding data and communications from the world. He makes an astonishing proposal to Thomas Stocker, a scientist at the Physics Institute of the University of Bern, Switzerland, in a 2009 e-mail:

> You might want to check with the IPCC Bureau. I've been told that IPCC is above national FOI Acts. One way to cover yourself and all those working in AR5 [the upcoming IPCC Fifth Assessment Report] would be to delete all e-mails at the end of the process. Hard to do, as not everybody will remember it.[80]

As an alarming part of the whole situation, the new e-mails revealed that the BBC, one of the world's leading news organizations, was heavily influenced by the Climatic Research Unit and the Tyndall Centre for Climate Research of the University of East Anglia. It appears that Mike Hulme, Professor of Climate Change at East Anglia, sponsored a seminar for BBC executives, named the Cambridge Media Environment Programme (CMEP). The CMEP received £15,000 over three years from UEA and delivered strong coaching to the BBC promoting man-made climate change. Hulme also vetted the four-part climate change BBC documentary *British Isles: A National History,* developed by Dan Tapster and released in 2004.[81]

See No, Hear No, Speak No Climate Evil!

"The e-mails are mainly about a controversy over a particular data set and the ways a particular small group of scientists have displayed that dataset."
—John Holdren, US Director of Science and Technology Policy[82]

In one e-mail, Hulme makes it clear that a goal of the UEA was to prevent coverage of scientists skeptical of man-made climate change. He remarks on a radio debate between John Houghton, a climate activist, and Philip Stott, a climate skeptic, in a 2002 e-mail:

> Did anyone hear Stott vs. Houghton on Today, radio 4 this morning? Woeful stuff really. This is one reason why Tyndall is sponsoring the Cambridge Media/Environment Programme to starve this type of reporting at source.[83]

With the help of the UEA, the BBC adopted a strongly biased position in favor of the theory of man-made global warming. After his retirement in 2009, veteran newsreader Peter Sissions described the bias of the BBC involving an interview in December 2008:

> On a wintery Saturday last December, there was what was billed as a major climate change rally in London. The leader of the Green Party, Caroline Lucas, went into the Westminster studio to be interviewed by me on BBC News channel. She clearly expected what I call a "free hit"; to be allowed to voice her views without being challenged on them. I pointed out to her that the climate didn't seem to be playing ball at the moment. We were having a particularly cold winter, even though carbon emissions were increasing. Indeed, there had been no warming for ten years, contradicting all the alarming computer predictions. Well she was outraged…Miss Lucas told me angrily that it was disgraceful that the BBC—the BBC!—should be giving any kind of publicity to those sort of views…But it was effectively BBC policy, enthusiastically carried out by the BBC environment correspondents, that those views should not be heard—witness the BBC statement last year that "BBC News currently takes that view that their reporting needs to be calibrated to take into account the scientific consensus that global warming is man-made."[84]

WHY DOES IT MATTER?

So why should we care that climate scientists at East Anglia University and others in the e-mail net were fudging the science? The answer is that the ramifications are global and affect every community on Earth.

Global temperature data sets are a mess. They are based on poor global coverage, with far fewer reporting sites than in past decades. US temperature station data, regarded as the best in the world, suffer from up to 2°C of error on over 70 percent of the sites. Scientific organizations have been caught adding warm temperature bias to the raw temperature data from Australia, New Zealand, Russia, and the United States.

The keepers of the temperature data, the Climatic Research Unit of East Anglia University, have had communications exposed for all the world to see. These communications show bias, manipulation of data, avoidance of freedom of information requests, and efforts to subvert the peer-review process, all to further the "cause" of man-made global warming. And the CRU tells us that they are unable to produce the original thermometer data on which the claims of unnatural warming are based.

A World Mislead by the IPCC

"…the findings for the IPCC's Fourth Assessment Report that…2050 targets for developed countries should be between 80 to 85 per cent below 1990 emissions."
—*Securing a Clean Energy Future*, Commonwealth of Australia (2011)[85]

"…IPCC reported that 'there is new and stronger evidence that most of the warming observed over the last 50 years is attributable to human activities.'"
—*Climate Change Scoping Plan*, California Air Resources Board (2008)[86]

"The IPCC's analysis suggests developed countries should collectively reduce their emissions by 25–40% below 1990 levels by 2020."
—*The UK Low Carbon Transition Plan*, Her Majesty's Government (2009)[87]

"…the UN IPCC…estimates that the average global surface temperature is likely to increase by between 1.4 and 5.8°C by 2100."
—*Climate Change Plan for Canada*, Government of Canada (2002)[88]

"…the IPCC has clearly indicated that most of the global warming observed over the past 50 years was likely induced by the increase in concentrations of greenhouse gases…"
—China National Climate Change Program (2007)[89]

"Most of the observed increase in global average surface temperatures since the mid-20th Century is very likely due to the observed increase in anthropogenic greenhouse gas concentrations.—IPCC"
—*Climate Change Indicators in the United States*, US Environmental
Protection Agency (2010)[90]

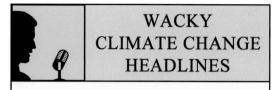

WACKY CLIMATE CHANGE HEADLINES

Study Says Global Warming Shrinks Birds
—*The Christian Science Monitor*,
Aug. 21, 2009[91]

Bigger Birds in Central California, Courtesy of Global Climate Change
—*SF State News*, Oct. 31, 2011[92]

The Climategate e-mail scientists, Phil Jones, Michael Mann, Keith Briffa, Jonathan Overpeck, Kevin Trenberth, and others, were all authors of the IPCC Assessment Reports of the Intergovernmental Panel on Climate Change. These IPCC reports have wrongly become the "gold standard" of climate science, quoted by all the governments of the world. The conclusions of the IPCC are the basis for misguided climate policies by every national government, state, province, town, and village across our earth. Cap-and-trade, carbon taxes, ethanol and biodiesel fuel mandates, renewable energy mandates, the banning of incandescent light bulbs, and many other misguided policies are the result. Incredible as it seems, this flawed scientific process has derailed the world over a trace gas in the atmosphere.

CHAPTER 10

FUND MY CLIMATE STUDY

"All modeling efforts will inevitably converge
on the result most likely to lead to further funding."
—JOURNALIST CHARLIE MARTIN

limate science has been corrupted. And industry has followed in tow. From university and government laboratories to renewable energy ventures and world financial markets, vast streams of money fuel the misguided ideology of Climatism. Deans of sustainability at universities, environmental vice presidents of corporations, environmental editors at newspapers, climate computer modelers, researchers of man-made climate "impacts," carbon traders, environmental groups, wind turbine manufacturers, solar cell manufacturers, and many others drink from a huge river of subsidies, tax breaks, grants, and consumer purchases generated from misguided global warming ideology.

THE SCIENTIFIC-GOVERNMENT COMPLEX

On January 17, 1961, President Dwight D. Eisenhower gave his farewell address to the nation. In his speech, he warned of two threats. The first was what he called the "military-industrial complex," which has become quite well-known in the media. But the second threat he described has unfortunately come to pass. He warned:

> Today, the solitary inventor, tinkering in his shop, has been overshadowed by task forces of scientists in laboratories and testing fields…Partly because of the huge costs involved, a government contract becomes virtually a substitute for intellectual curiosity…The prospect of domination of the nation's scholars by Federal employment, project allocations, and the power of money is ever present—and is gravely to be regarded. Yet, in holding scientific research and discovery in respect, as we should, we must also be alert to the equal and opposite danger that public policy could itself become the captive of a scientific-technological elite.[1]

Today, the world is in the grip of a scientific-government complex that carries the banner of Climatism. Over the last 20 years, government officials have been sold the false scientific doctrine of man-made global warming. The governments of the world have responded by feeding ever-larger amounts of cash to scientific organizations and universities to find solutions to the "problem" of man-made climate change.

An example of this scientific-government interplay is a 2008 position paper, titled "Advice to the New Administration and Congress: Actions to make our nation resilient to severe weather and climate change." The document was provided to both Barack Obama and John McCain, candidates at the time for the presidency. The paper was a plea for increased government funding for climate change research, written by US scientific organizations. It complained of a "consistent underfunding of these weather and climate research and operational programs" and asked for "9 billion beyond what our nation is planning to invest in this area between 2010 and 2014." The paper was issued by the University Corporation for Atmospheric Research, the American Meteorological Society, the American Geophysical Union, and others.[3]

YOU CAN'T HANDLE THE TRUTH!

It is no secret that a lot of climate-change research is subject to opinion, that climate models sometimes disagree…The problem is, only sensational exaggeration makes the kind of story that will get politicians'—and readers'—attention. So, yes, climate scientists might exaggerate, but in today's world, this is the only way to assure any political action and thus more federal financing to reduce the scientific uncertainty.

—Dr. Monika Kopacz, NOAA Program Manager (2009)[2]

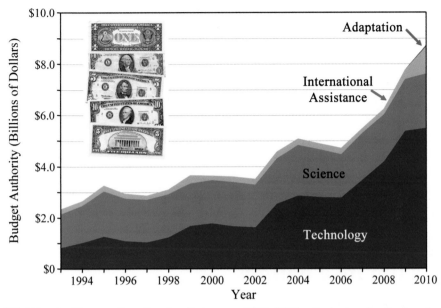

US Climate Change Funding by Category 1993–2010. Budget authority for climate change funding as reported by the Office of Management and Budget in four categories: technology, science, international assistance, and the newly-created category of adaptation. (GAO, 2011)[4]

Climate science is anything but underfunded. Over the last seventeen years, US government funding for climate change has exploded from $2.4 billion in 1993 to almost $9 billion in 2010. Among the largest blocks of federal funding in fiscal year 2010 were $4.6 billion to the Department of Energy (DOE), $1.2 billion to NASA, $567 million to the Department of Agriculture (USDA), $378 million to the Department of Commerce (primarily for NOAA), and $348 million to the National Science Foundation.[5] These organizations then provide research grants to scientific organizations such as the National Center for Atmospheric Research (NCAR) to study the impacts of man-made climate change.

The science portion of 2010 funding was $2.1 billion, a sixteen-fold increase from 1989 funding of $134 million.[6] Much of this money goes to computer modeling groups that generate alarming climate forecasts. Four main US organizations are conducting climate model

> ### Big Green for Climate Change
>
> **UI Gets Grant to Study Climate Change**
> "The University of Idaho announced the largest single grant in its history Friday, a $20 million award to study and plan for how climate change will affect cereal grain production in the Pacific Northwest."
> —*The Lewiston Morning Tribune*, Feb. 19, 2011[7]

simulations: the National Center for Atmospheric Research of the National Science Foundation, the Geophysical Fluid Dynamics Laboratory of NOAA, the Goddard Institute for Space Studies of NASA, and laboratories of the Department of Energy. Today's supercomputers cost over $50 million up front and require about $20 million per year to support a climate-modeling team.[8]

In 2008, the UK Meteorological Office installed a state-of-the-art IBM supercomputer, costing a tidy £33 million.[9] After Britain's record-cold December of 2010 that the Met Office failed to predict, Julia Slingo, chief scientist at the UK Meteorological Office, was asked how the Met Office could improve its forecasts. She replied:

Granthams to Fund Institute for Climate Change at Imperial College London
"Jeremy and Hannelore Grantham are donating GBP 12 million to establish the Grantham Institute for Climate Change based at Imperial College London..."
—London Imperial College,
    Feb 26, 2007[10]

Access to supercomputers. The science is well ahead of our ability to implement it. It's quite clear that if we could run our models at a higher resolution we could do a much better job—tomorrow—in terms of our seasonal and decadal predictions. It's so frustrating. We keep saying we need four times the computing power. We're talking just 10 or 20 million a year—dollars or pounds—which is tiny compared to the damage done by disasters. Yet it's a difficult argument to win.[11]

Oh, *now* we see. With just a lousy 10 or 20 million more each year, *then* you'll be able to forecast those weather disasters.

Purdue Wins $5M Global Warming Crop-Research Grant
"Purdue University scientists have won a $5 million federal grant to help corn and soybean farmers adapt to the various climate change scenarios global warming is forecast to bring in the coming decades."
—*Associated Press*, June 29, 2011[12]

It's educational to watch the Congressional testimony by officials from NOAA, NASA, the Center for Disease Control, and other governmental officials at hearings on climate change. No testimony is complete without a plea for increased funding so that "we can learn more about climate change."

Universities and colleges across the globe receive large amounts of money to study the impacts of man-made global warming. In the

United States, the DOE, NASA, NOAA, USDA, National Academy of Sciences, and EPA hand out grants in $1 million, $5 million, and $10 million dollar chunks to our learning institutions. If colleges want the funding, the one thing they must not say is "climate change is due to natural causes."

THE GREEN FLOW FROM FOUNDATIONS

Foundations are a major source of funding for Climatism. A survey by The Foundation Center estimated US foundation giving for climate change at $897 million in 2008, up nine-fold from $99 million in 2000. Leading organizations such as the Hewlett Foundation, the Packard Foundation, the Rockefeller Foundation, and the Kresge Foundation more than doubled their 2008 giving for activist climate programs.[13] *Politico* reported that US green groups spent about $500 million in 2009 to lobby for US climate legislation and to support global climate negotiations at the Copenhagen Climate Conference.[14] This was roughly balanced by $500 million in spending by oil, gas, and coal corporations in 2009 and 2010 to oppose cap-and-trade legislation.[15]

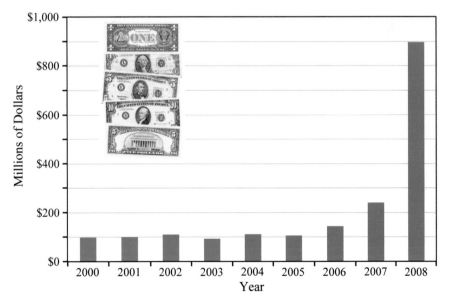

US Foundation Giving for Climate Change 2000–2008. Based on all grants of $10,000 or more awarded by a sample of over 1,000 of the largest US foundations. (The Foundation Center, 2008)[16]

FUNDING FOR AND FROM INDUSTRY

Corporations around the world are heavily invested in Climatism. General Electric (United States), Vestas (Denmark), Gamesa (Spain), Siemens (Germany), Goldwind (China), and many others sell tens of billions of dollars worth of wind turbine components and systems each year. Top solar cell manufacturers, such as Sharp Solar (Japan), Q-Cells (Germany), BP Solar, Shell Solar, and GE Solar, provide components for systems promoted by government climate policies. Archer Daniels Midland has built a big ethanol fuel business from the mandated blending of ethanol in US gasoline. All of these companies benefit from huge subsidies, feed-in tariffs, tax breaks and mandates from global governments. Total US national and state subsidies and tax breaks approach $20 billion per year.[18] In addition, the US Department of Energy provided over $35 billion in loan guarantees from 2009 to 2011 to industrial companies, primarily for renewable energy projects.[19]

ANOTHER GREEN ENERGY DEBACLE!

Solyndra Files for Bankruptcy, Looks for Buyer
"A California solar panel manufacturer that received more than a half-billion dollars in government loan guarantees filed for Chapter 11 bankruptcy…"
—*Business Week*, Sep. 6, 2011[17]

Carbon trading markets totalled $175 billion worldwide in 2011. BNP Paribas, Barclays Capital, and Deutsche Bank are leading players in the European carbon trading market, which was more than 80 percent of the global market. Wall Street firms Citigroup, Morgan Stanley, and Credit Suisse are big advocates for cap-and-trade and carbon trading in the US. Both financial institutions and renewable energy firms are large funders of activist climate policy.[20]

Even traditional hydrocarbon energy firms are big financial investors in Climatism. Exxon Mobil has been condemned for contributing millions to oppose climate legislation and to support skeptical science. But the company also donated $100 million of the $225 million in funding over ten years for the Global Climate and Energy Project at Stanford University. BP (formerly British Petroleum) donated $500 million for climate-change research at the University of California, Berkeley.[21]

And the funding continues. At his seventh annual Clinton Global Initiative meeting in 2011, former President Clinton announced commitments totaling $6.2 billion, much of it to fight climate change. Projects for "renewable energy, chickpea production, and promoting sustainable lifestyles" were funded by the initiative.[22]

THE MYTH OF WELL-FUNDED DENIERS

If the science is settled, then why haven't people been convinced that climate change is a problem that must be solved? Gallup announced results of a poll of more than 100 countries in 2011, finding that citizens in Australia, Canada, Europe, Japan, and the United States were less concerned about climate change than just a few years ago. When asked, "How serious of a threat is global warming to you and your family?," only 53 percent of Americans polled in 2010 answered "very concerned" or "somewhat concerned" vs. 63 percent when the same question was asked in 2007–2008. Concern dropped in Western Europe from 66 percent to 56 percent over the same period.[23] The proponents of Climatism argue that all people would accept the scientific consensus, except for uncertainty spread by "climate change deniers."

According to Dr. David Suzuki, a Canadian environmentalist:

> Despite the international scientific community's consensus on climate change, a small number of critics continue to deny that climate change exists or that humans are causing it. Widely known as climate change "skeptics" or "deniers," these individuals are generally not climate scientists and do not debate the science with the climate scientists…[24]

"Denial" is a venomous term applied to those skeptical about the role of mankind in global climate change, meant to equate climate change skeptics with Holocaust deniers. How can anyone seriously be opposed to saving the planet? True believers in Climatism can't understand how anyone can question the "overwhelming evidence" that man is causing global warming.

Dr. Gro Harlem Brundtland, environmentalist and former Prime Minister of Norway, judges that opposition to climate change activism is morally wrong:

> The diagnosis is clear, the science is unequivocal—it's completely immoral, even, to question now, on the basis of what we know, the reports that are out, to question the issue and to question whether we need to move forward at a much stronger pace as humankind to address the issues.[25]

PUNISH THE DENIERS!

"I wonder what sentences judges might hand down at future international criminal tribunals on those who will be partially but directly responsible for millions of deaths from starvation, famine, and disease in the decades ahead."
—Mark Lynas, UK journalist and environmentalist (2006)[26]

"When we've finally gotten serious about global warming, when the impacts are really hitting us and we're in a full worldwide scramble to minimize the damage, we should have war crimes trials for these bastards—some sort of climate Nuremburg."
—David Roberts, *Grist* (2006)[27]

So according to Ms. Brundtland, it's immoral *even to question* global warming dogma. Once judged as immoral, it's only a small step to condemn those who question the theory of man-made global warming as criminals.

A big myth, endlessly repeated, is that the "deniers" are well-funded. In an August 2007 *Newsweek* feature story, "The Truth About Denial," Sharon Begley describes what she calls a "denial machine":

> Since the late 1980s, this well-coordinated, well-funded campaign by contrarian scientists, free-market think tanks and industry has created a paralyzing fog of doubt around climate change. Through advertisements, opeds, lobbying, and media attention, greenhouse doubters (they hate being called deniers) argued first that the world is not warming, measurements indicating otherwise are flawed, they said. Then they claimed that any warming is natural, not caused by human activities. Now they contend that the looming warming will be miniscule and harmless. "They patterned what they did after the tobacco industry," says former senator Tim Wirth, who spearheaded environmental issues as an a undersecretary of state in the Clinton administration. "Both figured, sow enough doubt, call the science uncertain and in dispute. That's had a huge impact on both the public and Congress."[28]

According to Climatists, the scientific consensus is so strong that deniers *must be intellectually dishonest* and in the pay of oil companies. While signing books in bookstores, I've been asked on several occasions about which oil company is funding me. My wife also asks me when that first payment from an oil company is going to arrive.

In fact, a mountain of funding weighs heavily on the side of Climatism, including the $243 billion in annual global expenditures to shift to renewable energy, $20 billion in annual US tax breaks, $8 billion in annual US government funding, billions in government loan guarantees, major funding from foundations, and industry support. As additional evidence, compare the annual expenditures of environmental groups versus the "free-market think tanks" that Ms. Begley decries. The budgets of leading environmental groups are roughly a factor of ten larger than the think tanks. In all, the money promoting the theory of man-made warming is at least a factor of 20-50 times greater

PUNISH THE DENIERS!

"If you're one of those who have spent their lives undermining progressive climate legislation, bankrolling junk science, fueling spurious debates around false solutions, and cattle-prodding democratically-elected governments into submission, then hear this: We know who you are. We know where you live. We know where you work."
—Gene Ananth, Greenpeace, Apr. 1, 2010[29]

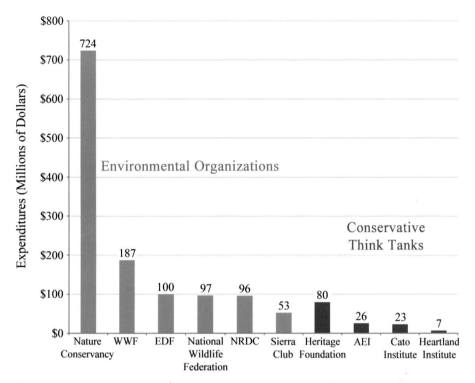

2010 Expenditures for US Environmental and Conservative Organizations. Expenditures for six leading environmental groups that promote activist climate change policy far exceed expenditures of conservative think tanks that oppose the theory of man-made global warming. Abbreviations are World Wildlife Fund (WWF), Environmental Defense Fund (EDF), Natural Resources Defense Council (NRDC), and American Enterprise Institute (AEI). Greenpeace not included. (2010)[30]

than the money funding skeptical opposition.

In fact, many of the leading climate skeptics (who should be called realists) are volunteers. Anthony Watts produced the audit of NCDC temperature stations as a volunteer leader of a group of volunteer auditors. Stephen McIntyre, who exposed the false science of the Mann Hockey Stick Curve, also did so as a volunteer. Watts remarks, "The theory was wrong, the evidence has changed, and thousands of volunteers have exposed it."[31]

THE GORE EFFECT

Gore Decries "Global Warming" in Bitterly Cold NYC

"With a near-record low temperature and single-digit wind chill in New York City, former Vice President Al Gore took to the podium in Manhattan's Beacon Theater today to blast President Bush for contributing to global warming."

—*World Net Daily*, Jan. 15, 2004[32]

IS MAN-MADE WARMING A HOAX?

Is man-made global warming a hoax? Senator James Inhofe, in a speech on the floor of the US Senate in 2003, stated:

> With all of the hysteria, all of the fear, all of the phony science, could it be that man-made global warming is the greatest hoax ever perpetrated on the American people? It sure sounds like it.[33]

Webster's New World Dictionary defines the term "hoax" as "a trick or a fraud, especially one meant as a practical joke." The term "hoax" implies that a person or a group of people have knowledge of the true situation, yet intentionally mislead or trick others.

I am astonished every day by an endless line of articles from all over the world, generated by scientists or news media, warning of the coming climate catastrophe. The overwhelming majority of these authors are sincere and strongly believe the Climatist ideology. Even the senders of the Climategate e-mails strongly believe that Earth is warming and that man is causing it. They apparently judge the situation to be so serious that cherry-picking of the data is warranted. Whether a hoax, a scam, or just a great mistake, global warming mania is a misguidedness unique in human history—truly a madness.

THE TAINTED INTEGRITY OF SCIENTIFIC ORGANIZATIONS

Despite mounting evidence to the contrary, the scientific organizations of the world remain solidly in the man-made global warming camp. The American Physical Society, the American Meteorological Society, NASA, NOAA, the National Academy of Sciences, the Royal Society of Britain, the European Academy of Sciences and Arts, and all other major scientific groups continue to espouse Climatism. Despite the fact that global temperatures have been flat for the last ten years, that dozens of peer-reviewed studies show that the Medieval Warm Period was warmer than today, that satellite and weather balloon data have not found the model-predicted warm spot over the tropics, that climate feedbacks appear to be low to negative, and that cosmic ray levels do appear to impact cloud formation, these groups remain unshakable in their support for activist climate policy. Nor has the discrediting of the Hockey Stick Curve, the exposure of IPCC bias, or the skullduggery of the Climategate e-mails seemed to have made any difference. Although these organizations officially support Climatist dogma, the members may not agree. Let's look at two of many examples: the American Meteorological Society and the American Physical Society.

Dr. William Gray of Colorado State University, a 50-year member of the American Meteorological Society (AMS), levels strong criticism at the organization:

I am very disappointed at the downward path the AMS has been following for the last 10-15 years in its advocacy of the Anthropogenic Global Warming (AGW) hypothesis. The society has officially taken a position many of us AMS members do not agree with…Instead of organizing meetings with free and open debates on the basic physics and the likelihood of AGW induced climate changes, the leaders of the society…have chosen to fully trust the climate models and deliberately avoid open debate and discussion…My interaction (over the years) with a broad segment of AMS members…have indicated that a majority of them do not agree that humans are the primary cause of global warming.[34]

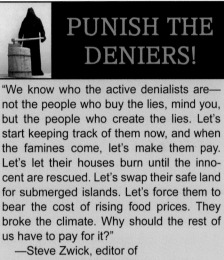

PUNISH THE DENIERS!

"We know who the active denialists are— not the people who buy the lies, mind you, but the people who create the lies. Let's start keeping track of them now, and when the famines come, let's make them pay. Let's let their houses burn until the innocent are rescued. Let's swap their safe land for submerged islands. Let's force them to bear the cost of rising food prices. They broke the climate. Why should the rest of us have to pay for it?"
—Steve Zwick, editor of
Ecosystem Marketplace,
Forbes, April 19, 2012[35]

In 2010, George Mason University conducted a survey of US television meteorologists who were members of the AMS and the National Weather Service, receiving responses from 571 weather forecasters. The results confirmed Dr. Gray's opinions. Almost two-thirds of the meteorologists (62 percent) judged that global warming was caused by natural changes in the environment, while 27 percent agreed with the statement "global warming is a scam."[36]

The leadership of the American Physical Society (APS) issued a 2007 statement supporting the theory of man-made global warming, stating "the evidence is incontrovertible." Several attempts by members of the APS to modify the statement have been fruitless. An open letter from 54 members of the APS in 2009 proposed that the organization adopt a more balanced position.[37] The letter resulted in a review by the APS Council, but no changes were made.

In 2010, distinguished physicist Harold Lewis, emeritus professor of the University of California, resigned from the APS. Just six months before his death, he penned scathing

Big Green for Climate Change

University of Florida-Led Teams Awarded $6.9 Million for Climate Change Projects
—*University of Florida News,*
June 30, 2011[38]

remarks in his resignation letter:

> When I joined the American Physical Society sixty-seven years ago it was much smaller, much gentler, and as yet uncorrupted by the money flood…the choice of physics as a profession was then a guarantor of a life of poverty and abstinence…How different it is now…the money flood has become the raison d'être of much physics research, the vital sustenance of much more, and it provides the support for untold numbers of professional jobs…It is of course, the global warming scam, with the (literally) trillions of dollars driving it, that has corrupted so many scientists, and has carried APS before it like a rogue wave. It is the greatest and most successful pseudoscientific fraud I have seen in my long life as a physicist.[39]

Similarly in 2011, Ivar Giaever, Nobel Prize winner in Physics in 1973, resigned from the APS, stating he could not live with the APS "incontrovertible" position on global warming:

> In the APS it is ok to discuss whether the mass of the proton changes over time and how a multi-universe behaves, but the evidence of global warming is incontrovertible? The claim (how can you measure the average temperature of the whole earth for a whole year?) is that the temperature has changed from ~288.0 to ~288.8 degree Kelvin in about 150 years, which (if true) means to me that the temperature has been amazingly stable, and both human health and happiness have definitely improved in the 'warming' period.[40]

When will the scientific organizations of the world show some intellectual integrity? Any honest climate scientists reading the Climategate e-mails should be sickened by what they see. But as Upton Sinclair reportedly said:

> It is difficult to get a man to understand something when his salary depends upon his not understanding it.[41]

CLIMATISM: PROPELLED BY MONEY

Governments jumped to a conclusion and adopted the theory of man-made global warming at the 1992 Rio de Janeiro Earth Summit. Today, Climatism is propelled forward by a huge tsunami of money. From the university research contract, to funding for the latest high-powered supercomputer, to building a multi-billion-dollar carbon trading market, climate alarmism is the ticket to growing your organization and climbing the ladder of fame and fortune. But as we'll see in the next chapter, the global obsession with renewable energy is the favored child of Climatism.

CHAPTER 11

SUNBEAMS, ZEPHYRS, AND GREEN, LEAFY FUEL

"To produce solar power in Germany is as sensible as to grow pineapples in Alaska."
—JUERGEN GROSSMAN, CEO OF ENERGY UTILITY RWE

Renewable energy is the answer, according to Climatism advocates. Many profess that renewables are affordable and at hand. They claim that if we switch to wind and solar energy, burn biofuels in our vehicles, and switch to electric cars, we can prevent the looming climate catastrophe.

Today, almost every government promotes the adoption of renewable energy. Many have been pushing renewables for more than 20 years. National, state, provincial, and local governments have used billions of dollars in subsidies, tax breaks, and loan guarantees, along with legislative mandates, to try to force a global transition away from hydrocarbon energy. Although energy security and other objectives are often stated as reasons, the misguided belief that climate change can be halted is at the bottom of all these programs. Yet, as movie detective Dirty Harry might say, renewables "ain't makin' it."[1] Let's look at the hard facts about renewable energy.

MODERN SOCIETY AND HYDROCARBON ENERGY

Modern society is built on energy. Energy use—and especially electricity, the most flexible form of energy—powers progress in all developed countries. A simple graph of income per person and energy use per person for various nations shows that energy use is strongly connected to prosperity.

From 1800 to 2010, world population increased seven-fold, from about one billion to about seven billion persons. But energy use increased even faster. World energy consumption in 2009 was estimated at 509 exajoules (billion-billion joules).[2] This energy use has increased by a factor of *more than 50* since 1800.[3] And hydrocarbon fuels supply the vast majority of this energy.

In 2010, about one billion cars and light trucks operated on roads across the world, with over 90 percent of their fuel coming from refined petroleum.[4] Ships, trains, and

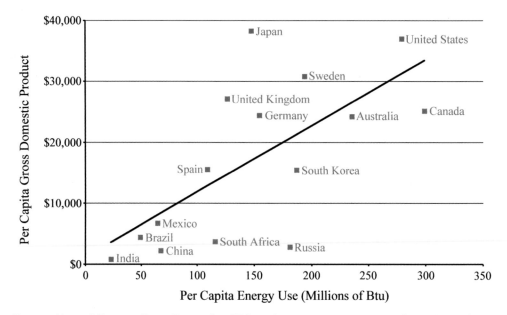

Prosperity and Energy Use. Per capita GDP and per person energy use for major nations. Countries that use more energy tend to also have higher levels of income. (Data from International Energy Agency 2011)[5]

planes are overwhelmingly powered by petroleum products. In 2009, 67 percent of the global electricity for lights, appliances, personal computers, cell phones, and machines was delivered by power plants using coal, natural gas, or oil.[6]

Prior to the 1800s, renewable energy sources were state of the art. Solar energy was captured by field and forest, providing forage for domesticated livestock. Animal muscle moved stagecoach and carriage. Wood was the primary fuel for heating and cooking, resulting in a growing deforestation of England, the Northeast United States, and other areas. Breezes were harnessed by windmills to grind corn and pump water and by sailing ships to cross the oceans.

During the last two centuries, hydrocarbon fuels replaced "medieval" energy sources for many important reasons. Coal was available for home heating, while wood resources were disappearing. But more importantly, burning hydrocarbons released the *concentrated* energy from the chemical bonds of these fuels. Coal contains double the energy density of wood.[7] Burning coal drove the steam engines of the Industrial Revolution. Pumps, industrial machines, locomotives, ships, and the first electrical power plants were powered by coal.

By the late 1800s, oil wells delivered petroleum in commercial quantities to industrializing nations. High-energy-density petroleum products such as gasoline, diesel, and kerosene quickly replaced coal as the fuels of choice for transportation. One ton of crude oil, when used as gasoline, releases about ten times as much energy as a ton of the explosive TNT.[8]

Natural gas was a cleaner fuel, free from the airborne particles produced by combustion of coal. Gas also had a high energy density and could be piped directly to residences and commercial buildings. Piping provided a major convenience advantage over oil and coal, which needed to be delivered by truck for small-volume applications. As a result, gas replaced oil and coal for building heating applications in the United States and other nations during the 1900s.

The "medieval" renewable energy sources of wind and solar, including wood and animal muscle, were replaced by hydrocarbons during the last two centuries. Hydrocarbons delivered more power, were less expensive, and could be counted on when needed.

SOME OPPOSE THE USE OF ENERGY

Many of today's intellectual elites view energy use as an evil of modern society. Some feel that mankind's increasing energy use is destroying the planet. Paul Ehrlich, Amory Lovins, Al Gore, and many others oppose increased energy use. Al Gore stated in his first book in 1992:

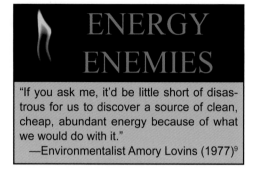

ENERGY ENEMIES

"If you ask me, it'd be little short of disastrous for us to discover a source of clean, cheap, abundant energy because of what we would do with it."
—Environmentalist Amory Lovins (1977)[9]

Our insatiable drive to rummage deep beneath the surface of the earth is a willful expansion of our dysfunctional civilization into Nature.[10]

The Club of Rome is just one of many organizations that oppose the use of energy. The group gained international recognition with publication of *Limits to Growth* in 1972, which advocated the "stopping of population growth in 1975 and industrial capital growth in 1985."[11] The Club has been a policy advisor to the United Nations for more than 30 years. At a 2011 conference, The Club proposed an "80 percent reduction in energy consumption."[12]

Energy opponents say that energy use must be limited and rationed if necessary. What better way to reduce energy growth than to embrace Climatism? If man-made emissions are destroying Earth's climate, then coal, oil, and gas usage must be stopped. Further, promotion of ineffective and costly renewable energy is an excellent method to retard global energy use.

RENEWABLE HYPE AND ENERGY REALITY

Despite the lessons of history and economic advantages of hydrocarbons, renewable energy has been promoted as a world tonic by media, universities, environmental groups and governments over the last 30 years. Renewables were first seriously touted as a solution to society's growing demand for oil products in the 1970s. With the rise of Climatism in the 1990s, renewable energy also became the answer to global warming, rising oceans, and polar bear extinction.

Predictions of renewable energy use have been hugely optimistic. In 1977, environmentalist Dennis Hayes, the founder of Earth Day, predicted:

> By the year 2000, such renewable energy sources could provide 40 percent of the global energy budget; by 2025, humanity could obtain 75 percent of its energy from solar resources.[13]

President Jimmy Carter's administration issued a report in 1980 titled *A New Prosperity, Building a Sustainable Energy Future*. This study laid out a plan that claimed to be able to provide 28 percent of United States energy needs from renewable sources by the year 2000.[14]

Renewables have made some progress, much of this by biofuels. The establishment of the Organization of Petroleum Exporting Countries (OPEC) in the 1960s, and the 1973 oil embargo, highlighted a growing dependence by the US and other industrialized nations on imports of oil from the Middle East. "Gasahol" was promoted by the US government in the 1970s as a replacement for gasoline. Ethanol and biodiesel incentives provided by the US and European Community (EC) fostered a growing

ANOTHER *HIGH-POTENTIAL* ENERGY SOURCE?

Senator Charles Schumer Touts Tax Incentives for Manure Energy
—*Associated Press*, Aug. 12, 2011[15]

global biofuels industry. World production of biofuels reached 105 billion liters (28 billion US gallons) in 2010, providing about 2.7 percent of all fuel for vehicle transportation.[16]

The first large-scale renewable systems for generating electricity were built in California the 1980s. Wind turbine farms were installed at Altamont Pass near San Francisco and at Tehachapi Pass and San Gorgonio Pass near Los Angeles, totalling 17,000 small turbines by 1990. The Solar Energy Generating Systems, nine facilities that were the world's first major commercial solar plants, were built in California's Mojave Desert in the 1980s.[17]

In the last two decades, major nations have built wind turbines in a frenzy. China, Germany, Spain, and the United States have each built tens of thousands of turbines. Denmark erected one turbine for every 1,000 citizens, the world's highest wind tower density. By the end of 2010, more than 160,000 wind turbines were operating across the world.[18]

Feed-in tariffs spurred installation of residential solar systems in many nations. A feed-in tariff requires a utility to buy electricity that is fed back into the grid from private solar systems, often at rates much higher than regular electricity prices. In the early 1990s, Germany established a generous feed-in tariff of over €50 cents per kilowatt-hour ($US 0.63/kW-hr), more than three times regular electricity rates. This resulted in more than one million residential solar installations in Germany during the last 20 years.[19] With the help of financial incentives, California residents and businesses installed over 50,000 roof-top solar systems.[20] France, Greece, Italy, and many other nations established strong financial or tax programs supporting solar energy. Even the United Kingdom, located at above 50°N latitude (not exactly the sun belt), enacted a solar feed-in tariff.

While all this renewable activity sounds impressive, it has hardly scratched the surface of the enormous level of world energy usage. According to the Energy Information Administration of the US Department of Energy

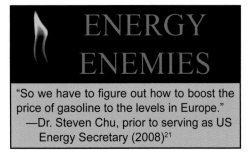

ENERGY ENEMIES

"So we have to figure out how to boost the price of gasoline to the levels in Europe."
—Dr. Steven Chu, prior to serving as US Energy Secretary (2008)[21]

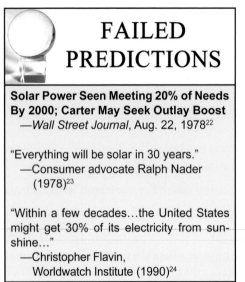

FAILED PREDICTIONS

Solar Power Seen Meeting 20% of Needs By 2000; Carter May Seek Outlay Boost
—*Wall Street Journal*, Aug. 22, 1978[22]

"Everything will be solar in 30 years."
—Consumer advocate Ralph Nader (1978)[23]

"Within a few decades...the United States might get 30% of its electricity from sunshine..."
—Christopher Flavin,
    Worldwatch Institute (1990)[24]

(DOE), world energy consumption is forecasted to grow about 1.6 percent per year through 2035.[25] A single year's growth adds about 200 million tons of oil equivalent to global energy usage, equal to adding a United Kingdom's worth of energy demand each year.[26]

Energy reality is that *human society overwhelmingly remains a hydrocarbon world.* In 1973, hydrocarbon fuels provided almost 87 percent of the world energy supply. By 2009, hydrocarbons still provided about 81 percent of the world supply. From 1973 to 2009, the share of world energy from oil declined from 46 percent to 33 percent, but the share from coal increased from 25 percent to 27 percent and the share from natural gas increased from 16 percent to 21 percent. Most of the reduction in hydrocarbon use was captured by increased use of nuclear power.[27]

Despite more than 20 years of hoopla and financial aid from governments, renewable energy remains an embarrassingly small part of the global energy picture. In 2009, vehicle biofuels provided only about 1 percent of global energy supply, only one tenth of the green "Biofuels and Waste" wedge in the diagram on the opposite page, which was mostly wood, charcoal, and dung combustion by poor nations. The contribution of wind and solar energy is located *within* the tiny red wedge labelled "Other." Renewables from solar, wind, and biofuels totaled less than 2 percent of world energy supply in 2009.[28]

Nor do statistics for recent years show much renewable progress. According to International Energy Agency data, hydrocarbons maintained their 81-percent share of world energy supply from 2006 to 2009.[29] Increased use of coal in China, India, and other developing nations offset global growth in renewables.

Renewable advocates say, "But look at the high growth rates for renewables!" However, this is growth upon a tiny base. Pushed by government incentives from 1995 through 2010, US electricity from wind increased by a factor of 30 and solar increased by a factor of 2.5. But during those same fifteen years, the increase in US electricity output from coal and natural gas totaled more than *six times* the increase in electricity from wind and solar. The increase in electrical output from natural gas was 637 times the increase in output

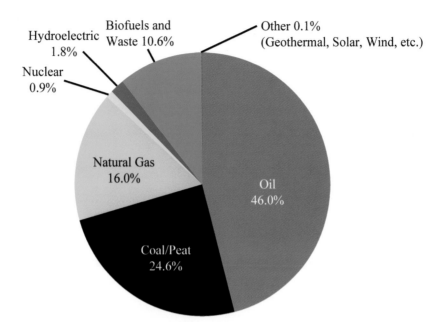

1973 World Energy Supply by Fuel. Biofuels and Waste includes wood and dung used by much of the developing world. (International Energy Agency, 2011)[30]

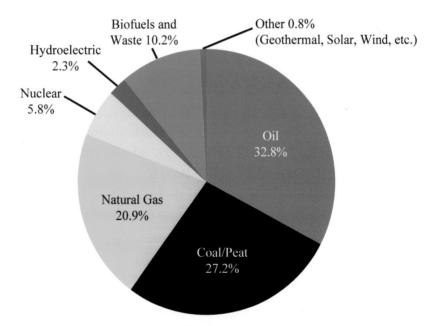

2009 World Energy Supply by Fuel. Biofuels and Waste includes wood and dung used by much of the developing world. (International Energy Agency, 2011)[31]

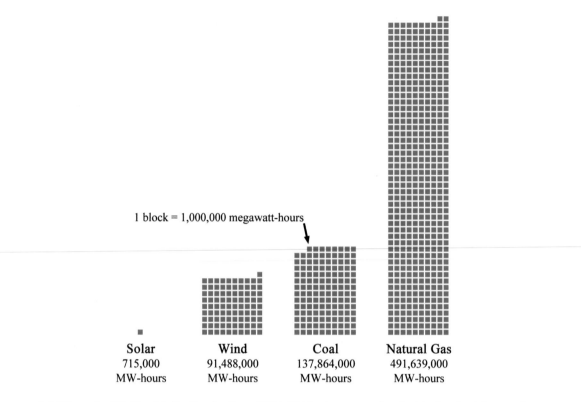

1 block = 1,000,000 megawatt-hours

Solar	Wind	Coal	Natural Gas
715,000	91,488,000	137,864,000	491,639,000
MW-hours	MW-hours	MW-hours	MW-hours

Additions to US Electricity Production 1995–2010. Increases in electrical output from solar, wind, coal, and natural gas over the last fifteen years in megawatt-hours of output. (Data from Energy Information Administration, 2012, after Bryce, 2010)[32]

from solar.[33] By any reasonable estimate, the amount of energy we get from solar remains trivial.

So why haven't renewables captured world markets? Consumer advocate Ralph Nader thinks a conspiracy is involved. When asked in 1992 why the US didn't have a national program to convert to solar energy, he replied, "Because Exxon doesn't own the sun."[34] But in fact, renewables have not been adopted despite the best efforts of promoting governments *because of the physics* involved. Renewable sources are dilute and intermittent. This makes them both less reliable and more costly than hydrocarbon or nuclear alternatives.

WIND AND SOLAR STRIKE ONE: DILUTE ENERGY

"All the energy stored in Earth's reserves of coal, oil, and natural gas is matched by the energy from just 20 days of sunshine." So says the website of the Union of Concerned Scientists.[35] Solar is a no-brainer, right?

The problem with this logic is that the surface of Earth is very large, and the sunlight that arrives at the surface at any one location is *dilute*. On a clear day when the sun is directly overhead, about 1,000 watts of sunshine reaches each square meter of Earth's surface at the equator after absorption and scattering by the atmosphere. For latitudes of Southern Europe and the US, this is reduced to about 800 watts per square meter at midday, since the angle of the sunlight is not quite perpendicular. The electrical conversion efficiencies of solar systems are about 10–20 percent. After accounting for power transmission losses, this means only a single 100-watt bulb can be powered for every card-table-sized surface area of a solar system, and only at noon on a clear day. For solar energy to try to replace conventional power plants, these "card tables" would have to multiply in a hurry.

Wind systems try to capture the similarly dilute energy of the wind. Wind turbines must be spaced about 140 meters apart,[36] so vast fields of wind turbines are required to capture a significant amount of energy. In addition, turbines must be able to handle the rapidly changing force of the wind. Wind energy varies with the cube of the wind speed. A 10-meter-per-second (m/s) wind generates 1,000 times the energy of a 1 m/s breeze. Wind turbines begin to operate at a light breeze of about 4 m/s but must shut down when wind speeds reach about 25 m/s.[37]

Another way to describe the dilute energy of wind and solar is to say that these sources have poor *energy density*. Energy density can be defined as the amount of power delivered per unit of land area occupied by the power plant. The lower the energy density of a system, the larger the area required to do the job. Let's look at five modern power plants to compare the land required.

The Kingsnorth coal-fired power plant is a proposed high-efficiency plant scheduled to begin operation in Kent, UK, in 2016. It will replace existing plants on the Kent site and, if constructed, will be the first coal-fired plant to be built in the UK since 1986. The design calls for an installed capacity of 1,600 megawatts (MW), with an average delivered power of 1,360 MW (85-percent capacity factor). Capacity factor is the percentage of rated power actually delivered on an annual basis. If built, the plant will occupy about 1.6 square kilometers

Burning Wind Turbine. This turbine in Germany failed to shut down in high winds.
—Photo by Polizei Stade[38]

(sq. km.) or about 0.6 square miles (sq. mi.) of land area.[39]

The planned Kingsnorth plant has been under withering fire from Greenpeace, World Wildlife Fund, and other groups for the last five years. The organizations claim that the plant will cause catastrophic global warming. Most recently, EON, the owning utility, has delayed the project and proposed adding carbon capture and storage (CCS) capability to reduce CO_2 emissions.[40]

Carbon capture and storage is a proposal to capture the emitted CO_2 from coal, gas, or industrial plants and store it underground. CCS has been proposed by both climate groups and coal companies as a solution to global warming. In short, it is a solution that will never happen due to high cost and the huge quantities of captured CO_2 that would need to be stored.

The second plant on our list is the Vogtle nuclear power plant planned to enter service in 2016 or 2017 near Augusta, Georgia. The plant consists of two reactors, together rated at 2,200 MW and estimated to deliver an average power of 1,980 MW, for a capacity factor of 90 percent. The facility will occupy 6.1 sq. km. (2.3 sq. mi.).[41]

The third plant for comparison is the West County natural gas facility located in Palm Beach, Florida. West County is actually three plants, two of which began operation during the last two years. The facility is rated at 3,750 MW and will operate at an 87-percent capacity factor, delivering 3,263 MW of average power. The plant is the largest hydrocarbon-fueled power plant in the US, but occupies a very small 0.9 sq. km. (0.3 sq. mi.) footprint.[42]

The Andosol-1 solar plant is our first renewable plant for comparison. Andosol-1 is called

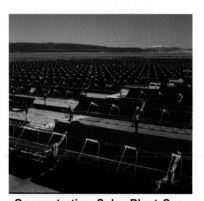

Concentrating Solar Plant System. Solar energy is reflected from parabolic troughs to a central pipe, which contains an absorbing fluid. (Photo by US BLM)[43]

a Concentrating Solar Plant (CSP). It uses a system of rows of parabolic mirrors that reflect the sun to heat a central pipe. The pipe contains a solution of molten salt, which captures the reflected solar energy. When built in Aldiere, Spain, in 2008, Andosol-1 was the third largest CSP plant in the world and the largest in Europe. The rated power of the system (also called nameplate power) is 50 MW. But because the system generates little power during morning and late afternoon hours, and no power at night, the system capacity factor is a low 16 percent, providing an average delivered power of only 8 MW. The system occupies 2 sq. km. (0.8 sq. mi.) in the Granada region in Spain.[44]

	Coal	Nuclear	Natural Gas	Solar	Wind
Name and Location	Kingsnorth Kent, UK	Vogtle 3 & 4 Augusta Georgia, US	West County Palm Beach Florida, US	Andosol-1 Aldiere, Spain	London Array North Sea, UK
Entered Service	Planned 2016	Planned 2016–2017	2009–2011	2008	2012
Installed Capacity (MW)	1,600	2,200	3,750	50	1,000
Average Delivered Power (MW)	1,360	1,980	3,263	8	370
Capacity Factor	85%	90%	87%	16%	37%
Site Footprint	1.6 sq km (0.6 sq mi)	6.1 sq km (2.3 sq mi)	0.9 sq km (0.3 sq mi)	2.0 sq km (0.8 sq mi)	245 sq km (95 sq mi)
Power Density (MW/sq km)	850	325	3,626	4	1.5

Power Plant Comparisons. Comparisons of modern coal, nuclear, natural gas, solar, and wind power plants. Note that coal, nuclear, and natural gas plants operate with significantly higher delivered power and capacity factors than solar and wind, while occupying much less land.[45]

The last system for comparison is the London Array offshore wind farm. The array is under construction in the North Sea off the east coast of England and scheduled to begin operation at the end of 2012. When its 341 wind turbine towers begin operation, it will be the largest offshore wind farm in the world, with a nameplate rating of 1,000 MW. Because the speed of the wind varies, the capacity factor of the London Array is estimated to be about 37 percent, providing 370 MW of average delivered power. The array will cover a huge 245 sq. km. (95 sq. mi.) area of the ocean.[46]

The table on the previous page summarizes the characteristics of these five world-class facilities, allowing us to compare and draw conclusions. First, the coal, gas, and nuclear plants all provide more than 1,000 MW of average delivered power, while the wind and solar facilities provide only 370 MW and 8 MW, respectively. Note that the Andasol-1 plant, among the largest solar plants in the world, provides a tiny average output of only 8 MW.

Second, the coal, gas, and nuclear facilities provide capacity factors from 85 to 90 percent. This means that electrical utility operators can count on these facilities to meet market demand when required. In contrast, the capacity factors for the London Array (37 percent) and Andosol-1 (16 percent) are low. Intermittency is the second major shortcoming of renewable energy sources, which we will expand on below.

Third, we can now compare the site footprint required for plants fueled by coal, nuclear, gas, wind, and the sun. To do this, we'll use the Kingsnorth coal plant as the standard. The average delivered power output of each plant can be scaled to Kingsnorth's 1,360 MW, and the resultant size of the footprint computed. After this calculation, the scaled conventional site sizes are 1.6 sq. km. (0.6 sq. mi.) for Kingsnorth coal, 4.2 sq. km. (1.6 sq. mi.) for Vogle nuclear, and 0.4 sq. km. (0.15 sq. mi.) for the West County gas plant. The site sizes for the coal, gas, and nuclear plants are 4.2 square kilometers (1.6 square miles) or smaller.

But when we scale the renewable facilities, we find that huge amounts of land area are required to deliver the same average power output. To deliver the same output as the Kingsnorth plant, Andasol-1 would need to be scaled up to 340 sq. km. (131 sq. mi.). This is equal to the metropolitan area of Atlanta, US, or Athens, Greece.[47] The London Array would need to be scaled up to 900 sq. km. (347 sq. mi.) to provide the same output as the Kingsnorth plant. This is five and one-half times the size of Washington, D.C. or *two-thirds the size of greater London.*[48]

So, to produce the same power output, a modern solar plant requires a land area between 75 and 100 times the size of a conventional power plant. A wind farm must cover an area 200 to 250 times as large. This analysis does not include the land area for mines that produce

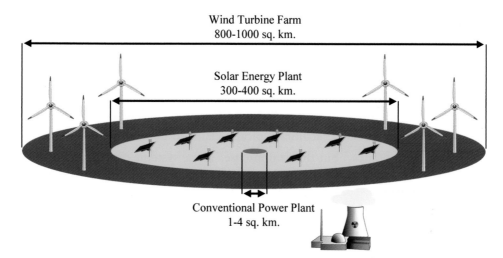

Surface Footprint Comparison of Energy Systems. Surface footprint of energy systems scaled to actual energy output. The conventional plants compared are planned Kingsnorth, UK, coal-fired plant; the planned Vogtle 3 and 4 nuclear plant, Augusta, GA; and the West County Energy Center gas-fired plant in Palm Beach, FL. The solar plant is Andasol-1 of Granada, Spain. The wind farm is London Array offshore system in North Sea east of the UK. Solar requires 75–100 times more area and wind requires 200–250 times more area than conventional power plants.[49]

the coal, gas, and uranium that provide fuel for conventional plants. Nor does it include the factories and mines that supply the huge amounts of material required for wind and solar structures. A single wind turbine requires 100–200 tons of steel and is anchored in a 1,000-ton steel-and-concrete base.[50] Wind advocates also argue that wind turbines allow alternative land uses while the wind farm is in operation.

Nevertheless, it's clear that conventional energy sources such as coal, gas, and nuclear have a huge energy density advantage over solar and wind. They provide reliable electrical power while using a small fraction of the land area of solar and wind. As we know, land isn't cheap in today's world. If two alternative power plants are being considered, and one occupies 75 times more land than the other, it's easy to predict which will be more cost effective.

By the way, how were solar and wind ever defined to be green and sustainable? How is it sustainable to cover thousands of acres of land with solar panels and wind turbine towers? All energy sources have disadvantages in terms of environmental impact, pollution, and cost. Common sense says that we should choose energy sources that provide the most power output while occupying the least land area, thereby having the smallest impact on the environment. Nuclear and hydrocarbon sources require a much smaller footprint than "green" solar and wind.

WIND AND SOLAR STRIKE TWO: INTERMITTENT ENERGY

The second big strike against renewable energy sources, and particularly wind and solar, is that they are *intermittent*. In 2010, the 36,000 wind turbines that were operating in the US delivered in total only 27 percent of their rated power output. As we just discussed, the Andasol-1 solar facility on average delivered only 16 percent of rated output. Announcements for new wind and solar facilities typically claim to be able to provide power for many "thousands of homes," but this is nonsense. These systems can't meet the needs of even a few homes on a 24-hour, 365-day basis.

Today's electrical systems are required to deliver power the instant it is demanded. When thousands of city dwellers return to their homes each evening and switch on their lights and air conditioners (or furnaces in the winter), the power must be available. Since the sun has already set or is low on the horizon, solar can't meet this demand. Without a breeze, wind can't meet the demand either.

Unlike the battery in your cell phone or flashlight, today's utilities have no method for large-scale electricity storage. Less than one percent of global electricity is stored for later delivery. Large-scale storage is technically possible in batteries or by other means but is cost prohibitive with today's technology.

Instead of using storage, electrical grid operators bring different power plants on line to meet demand as demand changes. Base-load power plants typically run 24 hours per day and deliver electricity at the lowest cost to the network. Nuclear and coal facilities usually serve as base-load plants because they are low cost and stable at full power. Intermediate, or load-following plants, are placed on and off line to meet changing demand. Gas turbine plants are often used in load-following service. Peaking-power plants are brought on line only during peak demand. Peaking-power plants are typically older coal or gas facilities that deliver the most expensive power, so they are used only as a last resort. Hydroelectric plants are flexible facilities that can be used in base-load, intermediate, or peaking applications.

Solar and wind energy don't fit into today's electrical power systems. They are unable to replace base-load power, since they are neither

Wind Turbines in Palm Springs, California.
(Photo by Turner, 2008)[51]

low cost nor capable of 24-hour operation. Nor are they able to serve in a load-following role. Summer peak demand is often in the evening, when solar output is low. Wind energy is just too variable to be counted on to meet peak demand.

For a real-world example, let's consider wind turbine performance as part of the National Grid of the United Kingdom. By the end of 2010, the UK had more than 3,300 wind turbines in operation. These turbines delivered about 2.6 percent of the nation's electricity, operating at a capacity factor of about 25 percent.[52] The wind system output during the winter of 2010–2011 is tracked by the red curve of the graph below. Wind output varied from a high of 100 percent to a low of almost 0 percent, with many periods where output was less than 20 percent of rated power. On December 7, when the system experienced peak electricity demand, National Grid reported that wind output was only about *5 percent* of rated power.[53]

Wind system output varies randomly and can change from full-rated power to zero output in only a matter of hours. Note that the graph shows output for all wind turbine towers in the United Kingdom. Wind conditions are often the same throughout the nation, and often the same throughout Northern Europe, resulting in region-wide shutdowns of wind turbines, depending upon weather conditions.

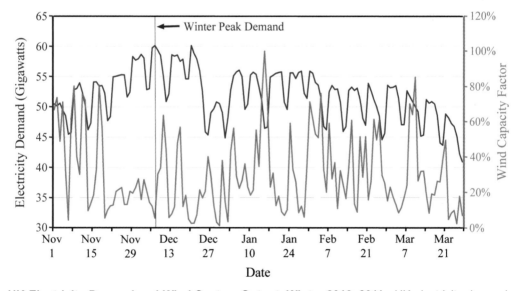

UK Electricity Demand and Wind System Output, Winter 2010–2011. UK electricity demand (blue curve) and wind turbine capacity factor (red curve). Note the huge variability of wind output and the frequent periods of less than 20 percent output, including at winter peak demand on December 7. Output was from more than 3,300 installed wind turbines operating at the end of 2010. (UK National Grid, 2011)[54]

ANOTHER *GREEN ENERGY* DEBACLE!

School to Create Own Wind Power
"A Cornish primary school could soon be almost completely powered by a single wind turbine. Gorran Primary School on the Roseland, has secured more than £50,000 from different agencies to carry out the work on the 15m (49ft) high turbine."
—*BBC News*, Feb. 29, 2008[55]

Wind Brings Down Turbine
"The eco-dream of a village school turned into a Friday 13th nightmare when high winds destroyed their wind turbine…Thank God it happened when the children were not out on the field."
—*New Quay Voice*, December 2, 2009[56]

That wind output was almost zero during the day of peak electricity demand is unfortunately not a rare occurrence. Days of high electricity demand tend to be either the coldest day of the winter or the hottest day of the summer. These extreme temperature days are usually caused by cold or hot high-pressure air masses, which are accompanied by clear skies and low wind conditions. This means that wind energy tends not to be available for times of peak demand.

Electrical grid systems are designed to meet peak demand. The sum of the reliable power output of all plants in a system must be greater than forecasted peak demand by an acceptable safety margin. Solar can be counted on if peak demand occurs in the afternoon, but only in regions where cloudy weather is rare. How much can wind systems contribute to meeting peak demand?

The UK National Grid estimates that 10 percent of wind is available for peak demand, although, as we have seen, only 5 percent was available for the 2010–2011 winter peak.[57] Texas has installed the largest wind turbine capacity of any US state. The grid operator ERCOT (Electric Reliability Council of Texas) "counts 8.7 percent of wind nameplate capacity as dependable capacity at peak," which is typically a summer day.[58] The Bonneville Power Administration (BPA) manages electricity for the northwestern United States. Peak load for BPA typically occurs on a winter day. Peak for 2008 occurred on December 16. At the time of peak demand, wind resources provided only about 7 percent of rated power.[60] A 2010 study by the Western Electric Council found that the "capacity value" of wind systems was only 10–15 percent of rated power.[61]

In 2008, a UK House of Lords study delivered a stunning conclusion about wind:

WISDOM FROM CLIMATISM?

"The grid is going to be a very different system in 2020, 2030. We keep thinking that we want it to be there and provide power when we need it…Families will have to get used to only using power when it was available, rather than constantly."
—Steve Holliday, UK National Grid CEO, Mar. 2, 2011[59]

But Doesn't Wind Energy Reduce Emissions?

The American Wind Energy Association states, "By directly reducing the use of fossil fuels, wind energy significantly reduces the emissions of the greenhouse gas carbon dioxide and other harmful pollutants."[62] But evidence is mounting that wind power *does not* significantly reduce greenhouse gas emissions.

Wind systems must have a coal or natural gas backup plant to assure continuity of electricity supply. Due to the frequent on-and-off nature of wind, backup plants must rapidly cycle between high output when the wind isn't blowing to low or zero output when the wind is blowing. This cycling is inefficient, resulting in greater emissions and greater fuel use from such back-up plants than would be the case with normal operation.

A 2010 study by Bentek Industries analyzed emissions from the Public Service Company in Colorado and the Electric Reliability Council of Texas for combined wind/hydrocarbon grid systems in both Colorado and Texas. Actual operating results showed that cycling of coal and natural gas plants to compensate for changes in wind output resulted in slightly more CO_2 emissions than if wind energy had not been part of the energy mix. More concerning, the study also showed an increase in total nitrogen oxide (NO_x) and sulfur dioxide (SO_2) emissions for each system, a direct result of the cycling of hydrocarbon plants.[63]

A 2009 study by Cornelius le Pair and Kees de Groot looked at the effect of adding wind energy to the electric grid system in the Netherlands. The study found that the addition of wind produced system inefficiency that caused "higher fuel consumption and CO_2 emission" than operation of the system without wind.[64] So even if man-made emissions were causing catastrophic global warming, the evidence shows that wind energy systems are not able to reduce such emissions.

The intermittent nature of wind turbines…means they can replace only a little of the capacity of fossil fuel and nuclear power plants if security of supply is to be maintained. Investment in renewable generation capacity will therefore largely be in addition to, rather than replacement for, the massive investment in fossil fuel and nuclear plant required…[65]

To repeat, *intermittency of wind means that this source can't replace hydrocarbon or nuclear-powered plants.* Further evidence is found in the examples of Germany and Denmark. By the end of 2010, Germany had constructed more than 20,000 wind turbines and Denmark more than 5,000. But neither nation has been able to close a single coal-fired power plant.[66]

WIND AND SOLAR STRIKE THREE: HIGH COST

Energy experts use "levelized cost" to compare electricity-generating costs from alternative energy sources. Levelized cost is the estimated cost of building and operating a power plant over the life of the plant, including all capital, fuel, maintenance, and fixed and variable costs at an assumed operating capacity factor. Estimates of levelized cost require many assumptions and should be therefore regarded as approximations of true costs.

As we just discussed, intermittent sources such as solar and wind cannot be compared in a direct way with hydrocarbon and nuclear, because they are not a true replacement for these sources. The US Department of Energy (DOE) draws this distinction:

> The duty cycle for intermittent renewable wind and solar is not operator controlled, but dependent upon the weather or solar cycle…as a result, their levelized costs are not directly comparable to those of other technologies…[67]

DOE cost estimates are graphed below for a facility put into operation in 2016. The estimates show natural gas to be the lowest cost at about 7 cents per kW-hr, with coal, hydro-electric, onshore wind, and nuclear costs between 9 and 12 cents per kW-hr. Offshore wind, solar photovoltaic, and solar thermal (like the Andasol plant) are at least double the price.[68]

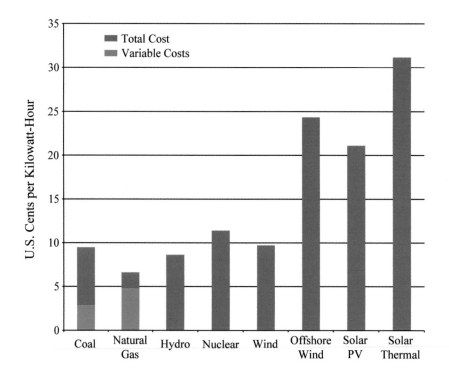

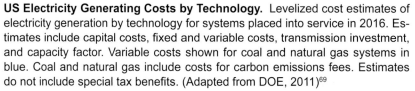

US Electricity Generating Costs by Technology. Levelized cost estimates of electricity generation by technology for systems placed into service in 2016. Estimates include capital costs, fixed and variable costs, transmission investment, and capacity factor. Variable costs shown for coal and natural gas systems in blue. Coal and natural gas include costs for carbon emissions fees. Estimates do not include special tax benefits. (Adapted from DOE, 2011)[69]

But a closer look at how these costs were arrived at is illuminating. First, DOE artificially added a "carbon dioxide emissions fee" to coal and natural gas generation, raising the levelized costs of these fuels. Second, DOE provided a very favorable estimated cost for onshore wind. A 34-percent capacity factor was assumed in the calculations,[70] better than the 2010 national wind average capacity factor of 27 percent and even the 30 percent average of new installed systems in 2011.[71] Just two years ago, the 2009 DOE estimated the levelized cost of onshore wind at fourteen cents per kW-hr.[72] Since US wind turbine installation costs have been rising for several years, it's surprising that DOE reduced their cost estimate for wind in the 2011 analysis by over 30 percent.

Climate Hypocrisy?

**Al Gore's "Inconvenient Truth"?
—A $30,000 Utility Bill**
"…electric bills for the former vice president's 20-room home and pool house devoured nearly 221,000 kilowatt-hours in 2006, more than 20 times the national average of 10,656 kilowatt-hours."
—*ABC News*, Feb. 26, 2007[73]

ENERGY ENEMIES

"Giving society cheap, abundant energy would be the equivalent of giving an idiot child a machine gun."
—Paul Ehrlich (1970)[74]

As we stated already, wind and solar can't replace a conventional power plant because of the intermittent nature of their energy. Since the conventional plant must be in place to meet peak electricity demand, the wind and solar facility can only reduce the utilization of a conventional plant. Since these renewable sources are only substitutes for part of the variable operation of a conventional plant, the cost of wind and solar should be compared to the *variable cost* of energy production by coal, natural gas, hydroelectric, and nuclear. As shown in the US electricity generating costs diagram, the variable cost of coal and natural gas is 2–5 cents per kW-hr, much lower than the cost of wind and solar substitutes.

Renewable proponents often argue that wind and solar energy are coming down a steep learning curve. They claim that if we only maintain the government subsidies and mandates for a few years longer, these energy sources will be competitive with traditional fuels. But the data for wind tells a different story. The installed cost of wind energy declined for many years, but reached a low point about 2002. Since 2002, the installed cost of US wind turbines has climbed 65 percent from $1,300 to about $2,150 per kW. The average installation of a 2 MW wind turbine now costs about $4.3 million in the US.[75] With more than 160,000 wind turbines installed worldwide, wind technology is already well down the learning curve, so rapid further cost reductions are unlikely.

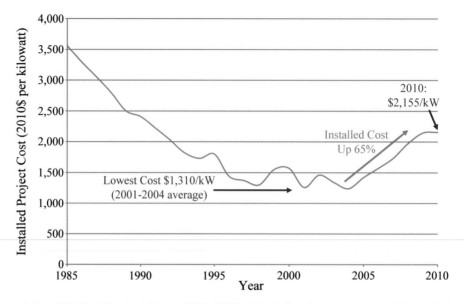

Installed US Wind Project Costs 1985–2010. After falling for many years, US installed wind turbine costs have increased 65 percent during the last decade. The average 2-megawatt wind turbine now costs about $4.3 million. (Data from DOE, 2011)[76]

In fact, wind costs are likely to rise. Opposition is growing to onshore wind in many countries. Denmark, the Netherlands, the United Kingdom, and other nations are turning their focus to offshore wind because of local opposition and a shortage of new good onshore locations. Since offshore wind is about double the price of onshore wind, expect wind installation costs to continue to rise.

Regarding solar, advocates have long claimed that the cost of solar cells will drop steeply, improving the competitive cost of solar energy systems. The price of solar cells has dropped more than 50 percent over the last few years, but solar systems are still not competitive. Remember that solar systems require about 75 times the land of conventional power plants. *Even if the cost of solar cells goes to zero*, the cost of land, holding structures, power lines, and facilities will still be higher than alternative conventional systems.

WACKY CLIMATE CHANGE HEADLINES

Wind Farm Revolts Blamed for Dramatic Fall in Planning Approvals
"Local revolts by British residents against wind farms are to blame for the number of planning approvals hitting record lows…"
—*Telegraph*, Oct. 28, 2010[77]

THE BIOFUELS FIASCO

The zenith of climate change madness is the global drive for biofuels. Originally a solution to reduce imports of Middle East oil, biofuels have been embraced by the United States and the EC as a solution to reduce emissions from modern transportation systems. Biofuels primarily consist of ethanol, which is blended with gasoline, and biodiesel, which is blended with diesel fuel, as partial substitutes for petroleum-based fuels. Ethanol is produced primarily from corn, sugar cane, wheat, or barley. Biodiesel is produced from the oil of rapeseed, sunflower, soybean, palm and other plants, or animal fat.

For 40 years, the world's political leaders have touted the benefits of biofuels. The Energy Tax Act of 1978 established the US "gasohol" industry, providing a subsidy of 40 cents per gallon for ethanol blended with gasoline.[78] President George W. Bush promoted the development of a US biofuels industry to reduce US dependence on foreign oil. In a 2006 speech, he gushed about ethanol fuel:

> I set a goal to replace oil from around the world. The best way and the fastest way to do so is to expand the use of ethanol…the use of ethanol in automobiles is good for the agricultural sector…Ethanol is good for rural communities…Ethanol is good for the environment… And ethanol's good for drivers. Ethanol is home-grown. Ethanol will replace gasoline consumption. Ethanol's good for the whole country.[79]

British Prime Minister Tony Blair was a champion of biofuels as a way to fight climate change. He wrote the foreword for the 2003 UK government report, titled *Our Energy Future—Creating A Low Carbon Economy*. The report called for biofuels to capture five percent of UK vehicle fuel by 2020, a percentage goal that would be raised throughout the decade.[80]

German Chancellor Angela Merkel has been a strong advocate of biofuel as a tool to reduce climate change. In a 2007 speech she stated:

> I am convinced that one of the greatest challenges of our time is climate change…Today we declared that we want to work together in several key areas related to climate change and energy. Therefore we have developed a common framework—which includes biofuels…The Americans have an ambitious goal of introducing 20 percent biofuels. The European Union has set itself the goal of 15 percent. We can collaborate here.[81]

Convinced of the need to fight global warming and convinced that biofuels were an answer to reducing emissions from vehicles, governments established incentives and mandates that resulted in a massive biofuel production program. The US Energy Independence and

Security Acts of 2005 and 2007 established a Renewable Fuel Standard (RFS) that imposed mandates on US oil refineries. The RFS required the blending of 11.1 billion gallons of biofuels into fuels in 2009, rising every year to a required 36 billion gallons in year 2022. Conventional biofuels, such as those made from corn, were capped at a maximum of 15 billion gallons. Second-generation biofuels, cellulosic fuels made from wood chips or switch grass, are intended to provide the balance of the biofuel totals during later years.[82]

The EC also set biofuel goals. The 2003 Biofuels Directive established a goal of replacing 2 percent of vehicle fuel by 2005 and 5.75 percent by the year 2010, with the objectives of:

> …meeting climate change commitments, environmentally friendly security of supply and promoting renewable energy sources.[83]

In 2007 the EC boosted the target to 10 percent by 2020.[84] The United Kingdom, Germany, and other European nations adopted these goals.

Across the world, biofuels are heavily subsidized at all levels of government. A 2007 study by the International Institute of Sustainable Development estimated total ethanol subsidies in the range of $1.00–1.25 per gallon ($0.26–0.33 per litre) and biodiesel subsidies in the range of $0.75–2.50 per gallon ($0.20–0.66 per litre).[85] Support per unit was highest in Europe, but total subsidies paid were the highest in the US. By 2010, global biofuel subsidies exceeded $15 billion per year.

These mandates and accompanying subsidies resulted in an explosion in world biofuel production. Global output increased from about 100 billion barrels in 2000 to almost 700 billion barrels in 2010.[86] The US is the world's largest producer and consumer of ethanol, with Brazil a noted second large producer. Europe is the leading producer and consumer of biodiesel fuel. US passenger cars run on a blend of 90 percent gasoline, 10 percent ethanol and also E85, a fuel using 85 percent ethanol. Brazil uses E25, a blend of 25 percent ethanol. Diesel fuel comes as B2, B5, B20, and B100, with the number indicating the percentage of biodiesel blended.[88]

So the world has established a biofuel industry. Vehicles are now being run on green, leafy biofuels. But a closer look shows that biofuels are neither good economics nor good environmental policy.

When you consider the energy physics

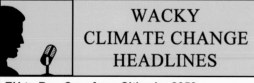

WACKY CLIMATE CHANGE HEADLINES

EU to Ban Cars from Cities by 2050
"Cars will be banned from London and all other cities across Europe under a draconian EU master plan to cut CO$_2$ emissions by 60 per cent over the next 40 years.
—*Telegraph*, Mar. 28, 2011[87]

of biofuels, they don't compare well with hydrocarbon fuels. Ethanol isn't a good substitute for gasoline. First, the energy content of ethanol is 76,000 British Thermal Units (Btu) per gallon, only about 66 percent of the 115,000 Btu content of gasoline.[89] This means that a vehicle using ethanol will get only two-thirds

ANOTHER HIGH-POTENTIAL ENERGY SOURCE?

Alligator Fat as a New Source of Biodiesel Fuel
—*Bio Fuel Daily*, Oct. 28, 2011[90]

of the mileage of a gasoline-powered vehicle. This isn't noticeable when using a 10 percent ethanol blend. But a car that gets 30 miles per gallon (mpg) on pure gasoline will only get 21.4 mpg when it runs on E85.

Second, energy is required to process plant feedstocks into ethanol or biodiesel. This includes the energy used to produce seed, fertilizer, irrigation, and insecticides for agriculture. Fuel is required for harvesting equipment and for transportation to the biofuel processing plant. The processing plant requires electricity and natural gas to refine the biofuel. Petroleum refining also requires energy, but the energy used in biofuel production is much higher on a per-barrel basis. Estimates vary regarding the energy gain from producing ethanol, but the US Department of Agriculture (USDA) places the net gain at only 34 percent.[91]

These two factors severely diminish the value of using corn ethanol as a replacement for gasoline. If we take the USDA estimated energy gain of 34 percent and combine it with ethanol's energy density of 66 percent of gasoline, we compute a net energy gain of only 22 percent for ethanol used as a substitute for gasoline. This means that, on an energy equivalent basis, *four and one-half gallons of ethanol from corn must be produced to replace one gallon of gasoline.*

Ethanol from sugar cane and biodiesel have better energy metrics. Ethanol from sugar cane produces about 3.7 units of energy for each unit of energy expended in production of the cane-ethanol.[92] But sugar cane ethanol still gets only two-thirds of the mileage of gasoline. Also, sugar cane can be grown in Brazil, but not cost-effectively in most of the United States.

Biodiesel delivers better energy density than ethanol. Biodiesel has an energy content of about 118,000 Btu, 3 percent more than gasoline, but less than the 130,000 Btu content of pure diesel fuel.[93] Therefore, the mileage of biodiesel is as good or better than pure gasoline. In addition, for each unit of energy used to grow and refine biodiesel, about 3.5 units of energy are created.[94] The big problem with biodiesel is the low yield per acre of biodiesel crops. US corn crops provide about 375 gallons of ethanol per acre compared to only about 52 gallons of biodiesel per acre from soybeans.[95] With the possible exception

IS THIS WHAT YOU MEAN BY SUSTAINABLE???

Texas Windfarm (US DOE)[96]

of ethanol from Brazil sugar cane, the physics of energy density and required energy input do not favor biofuels over gasoline or diesel fuel produced from petroleum.

Like wind and solar, biofuels are poor examples of sustainability. Energy density from harvested crops is low. In addition to the land footprint, biofuel production requires huge amounts of fertilizer, pesticides, and water.

The low energy density of biofuels means that the amount of land required to provide a significant amount of fuel is *enormous*. The Food and Agriculture Organization of the United Nations (FAO) made this shocking statement about biofuels in 2006:

> It is estimated that the EU would need to convert about 70% of its agricultural land to provide 10% of its energy need.[97]

That needs to be repeated. To provide 10 percent of its vehicle fuel from biofuels, the European Union would need to use *70 percent of its agricultural land*.

Again we run into the sheer magnitude of world energy usage. In 2010, the world produced almost 700 million barrels of biofuels, but this was less than 3 percent of world vehicle fuel usage of about 25 billion barrels.[98] To provide just 10 percent of world vehicle fuel from biofuels, nation-sized amounts of land are needed.

The results of the ethanol explosion are apparent in the United States. In 2001, just 7 percent of US corn production was used for ethanol fuel. By 2010, ethanol fuel was taking 34 percent of the harvest, but providing just 5.5 percent of US motor vehicle fuel.[99] Future mandates, if not rescinded, will force more than half of the US corn crop to be used for fuel.

Clairvoyance isn't needed to guess that, if we use farmland to grow both food *and* vehicle fuels, the price of food will rise. After decades of declining real prices, world prices for corn and soybeans doubled from 2002 to 2010. These increases drove global price increases for vegetable oils, cereals, and dairy products. A 2008 FAO report on biofuels found:

> Rapidly growing demand for biofuel feedstocks has contributed to higher food prices, which pose an immediate threat to the food security of poor net food buyers (in value terms) in both urban and rural areas.[100]

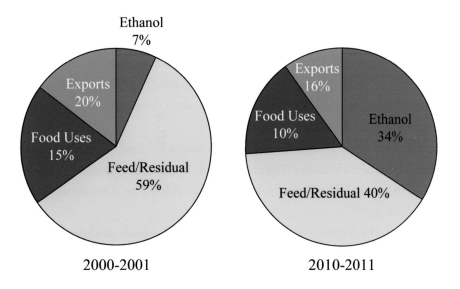

Uses of US Corn 2000–2001 and 2010–2011. Ethanol use for fuel is now more than one-third of US corn production. (USDA, 2011)[101]

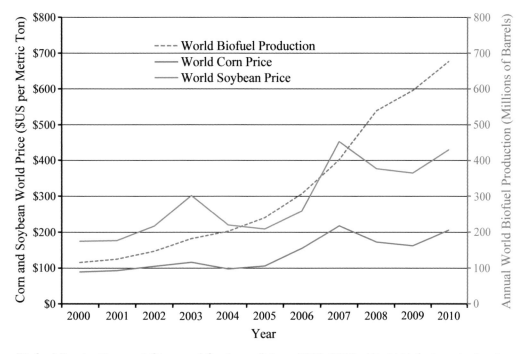

Biofuel Production and Corn and Soybean Prices 2000–2010. World biofuel production has increased by a factor of seven over the last ten years. Corn and soybean prices have doubled over the same period. (US Dept. of Energy, Food and Policy Research Institute, 2011)[102]

Donald Mitchell of the World Bank estimated that 70–75 percent of the price increase in food commodities from 2002 to 2008 was due to biofuel production and associated policies.[103]

Each bushel of corn produces about 2.7 gallons of ethanol. A sport utility vehicle using E85 fuel consumes about 25 gallons of ethanol per tank. Therefore, a single tank of E85 uses over nine bushels of corn, which according to some estimates can feed a person in the developing world for a full year.[104]

Do sustainable energy sources use less water? Not in the case of biofuels. A 2009 study by Argonne National Laboratory found that even in best-case agricultural conditions in US midwestern states, ethanol production requires from 2 to 3 times more water than gasoline production.[105]

Researchers at the University of Twente in the Netherlands analyzed water usage for twelve different biofuel crops in 2009. They found that, as a global average, ethanol production requires about *40 times* more water than gasoline production, or about 268 gallons of water used for each gallon of ethanol produced. Biodiesel production from soybeans uses about *268 times* more water than gasoline production, a whopping 1,968 gallons of water used for each gallon of biodiesel fuel produced.[106] Is that sustainable? Imagine how much additional water we could drink up if all vehicle fuel comes from green, leafy biofuels.

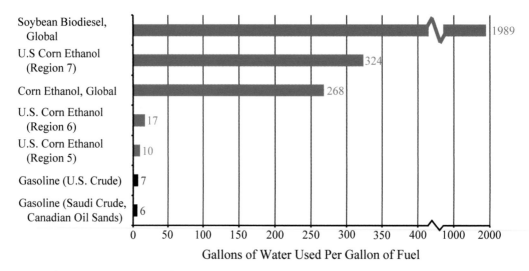

Water Consumption for Biofuels and Petroleum Gasoline. Gallons of water consumed per gallon of fuel produced for gasoline, ethanol, and biodiesel from various sources, including irrigation and fuel production, but not including precipitation. Variations in water consumption for three US regions and global averages for ethanol and biodiesel are primarily due to amount of irrigation used and agricultural yield. (Argonne National Laboratory, 2009; University of Twente, 2009)[107,108]

With the questionable "benefits" of poor gas mileage, huge land use, higher food prices, excessive water consumption, and required subsidies, biofuels must deliver powerful gains in pollution and greenhouse gas reductions, right? Well, actually no. A 2011 study for the National Academy of Sciences found that:

Climate Oxymoron?

"Sustainable Biofuels"

…production and use of ethanol as fuel to displace gasoline is likely to increase such air pollutants as particulate matter, ozone, and sulfur oxides.

The study also found that greenhouse gas emissions from corn ethanol are generally *higher* than emissions from gasoline.[109] Other recent studies show that biofuel use degrades air quality and also fails to reduce greenhouse gas emissions.

As a result, many environmental groups, including Greenpeace, Friends of the Earth, and Oxfam, have recently changed from biofuel advocacy to opposition. Even former Vice President Gore changed his mind on corn ethanol. In 1994, he cast the deciding vote in favor

---

The Biofuel "Greenhouse Gas Accounting Error"

For many years, the proponents of Climatism assumed that the burning of biomass would be inherently "carbon neutral." If true, the substitution of biofuels for hydrocarbon vehicle fuel would reduce global greenhouse gas (GHG) emissions. This is the basis for many climate change policies in the US, Europe, and other areas.

But the burning of wood or plant material releases CO_2 into the atmosphere, like any other combustion. Burning 1 metric ton of dry wood releases about 1.8 metric tons of CO_2 into the atmosphere.[110] In addition, since biomass has lower energy density than coal, gas, or petroleum products, direct burning of biomass releases *more* CO_2 than hydrocarbon fuels.

The "carbon neutral" concept assumed that as biofuel plants grow they absorb CO_2 equal to the amount released when burned. But, as many studies have recently pointed out, this GHG accounting does not consider vegetation that would naturally grow on land not used for biofuel feedstock production. Foliage grows naturally on land not used for agriculture, absorbing CO_2 from the atmosphere. Increasing global biofuel production would use an increasing amount of land for feedstock, removing this land from natural plant growth and sequestration of carbon.

A 2011 opinion from the Science Committee of the European Environment Agency calls this mistake a "serious accounting error."[111] Since biofuels are less efficient than gasoline or diesel fuel, they actually emit more CO_2 per gallon than hydrocarbon fuels, when proper accounting is used for carbon sequestered in natural vegetation. World governments are *actually boosting emissions* with biofuel policies.

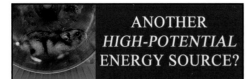

ANOTHER *HIGH-POTENTIAL* ENERGY SOURCE?

Obama's Energy Plan—Algae
"We're making new investments in the development of gasoline and diesel and jet fuel that's actually made from a plant-like substance—algae...we could replace up to 17 percent of the oil we import for transportation with this fuel that we can grow right here in America."
—President Barack Obama, Feb. 24, 2012[112]

of ethanol mandates to break a 50–50 US Senate deadlock. But in 2010, he stated, "It is not a good policy to have these massive subsidies for first-generation ethanol."[113]

Let's tally the scorecard. Biofuels get lower mileage than gasoline or diesel fuel. Biofuel production uses from twice as much to dozens of times more water than petroleum fuels. Biofuels produce more SO_2, NO_x, and ozone air pollution. They cause world grain prices to rise, adding to starvation in developing nations. They require many times more land than petroleum alternatives. Biofuels are more expensive than petroleum, requiring large subsidies and mandates from governments. Finally, when you look at the whole life cycle of biofuels versus petroleum vehicle fuels, burning biofuels even produces more CO_2 emissions.

Nevertheless, Climatist ideology that "biofuels reduce global warming" lives on. The EC remains unshaken in its directive of "10% share of sustainably produced biofuels" by 2020 for member nations.[114] The US Environmental Protection Agency is promoting E15, a 15 percent ethanol blend for light cars and trucks, stating:

> Increased use of renewable fuels in the United States can reduce dependence upon foreign sources of crude oil and foster development of domestic energy sources, while at the same time providing important reductions in greenhouse gas emissions that contribute to climate change.[115]

President Obama issued an executive order in May 2011 ordering that all US federal government cars and light trucks be "alternative fueled" by 2015.[116] This is a mandate to convert the government vehicle fleet to electric, hybrid, or biofuel alternatives. Which brings us to electric cars, our next climate madness topic.

ELECTRIC CARS—"DRIVEN" BY CLIMATISM

Electric cars are not really a new idea. Back in the early 1900s, at the dawn of the age of automobiles, plug-in electric vehicles held the majority of the infant US car market. But as the industry developed, the power of gasoline- and diesel-powered vehicles soon left electrics in the dust. Improving road systems connected cities, supporting a need for longer-range vehicles. The growing US oil industry reduced the price of gasoline, making it affordable to the average consumer. The invention of the electric starter by Charles Kettering in 1912 eliminated the hand crank starter, which was previously a disadvantage for gasoline cars. Finally, mass production of internal combustion engine vehicles by Henry Ford dropped their price to about one-third of the price of electric cars. Electric cars all but disappeared from the US market by the mid-1930s.[117]

Thomas Edison and an electric car in 1913. (Smithsonian)[118]

The twenty-first century revival of electric cars is driven by Climatism. World leader after world leader has embraced plug-in hybrid electric vehicles (PHEVs) as a solution for reducing emissions. Canada, China, nations of Europe, Japan, and the United States are providing generous incentives such as purchase payments to consumers, subsidies and loan guarantees to manufacturers, and public programs to install battery charging stations.

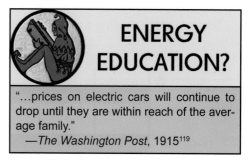

ENERGY EDUCATION?

"…prices on electric cars will continue to drop until they are within reach of the average family."
—*The Washington Post*, 1915[119]

President Obama has pledged to put one million PHEVs on the road in the US by year 2015. His proposed 2012 budget called for boosting the federal tax rebate from $7,500 to $10,000 for purchase of a PHEV.[120] The US Environmental Protection Agency and US Department of Transportation adopted a 54.5 mpg Corporate Average Fuel Economy Standard for year 2025, with the intention of forcing consumer adoption of PHEVs.[121]

Electric cars do have some advantages. Electric car engines are quieter than internal combustion engines. Electric engines are comparatively simple, reducing maintenance costs. Electric engines are more efficient than internal combustion engines. Finally, electric cars have "zero emissions," which is the fundamental reason for the misguided public push.

But plug-in electrics still suffer from the same major disadvantages that caused their disappearance a century ago. These disadvantages stem from the fact that electric cars get their energy from *batteries*. The physics and chemistry of batteries do not match the power of an internal combustion engine using gasoline or diesel fuel. Disadvantages include short driving range, long charging times, cost, and short battery lifetime.

First, today's best batteries can hold only a tiny fraction of the energy in the chemical bonds of gasoline. The University of Tennessee Center for Energy estimates the energy density of today's lithium-ion batteries at about 120 watt-hr. per kilogram.[122] Industry estimates place gasoline and diesel fuel at more than 12,000 watt-hr. per kilogram, an advantage of about a factor of 100.[123] Electric motors are about four times as efficient, but this still means that internal combustion engines have a power advantage of 20 to 25. Electric car manufacturers must include large, heavy battery packs just to go short distances, making electric cars heavier than gasoline counterparts. Typical gasoline-powered cars have a range of more than 350 miles before refueling. Today's electric car leaders, the General Motors Volt® and the Nissan Leaf,® have electric ranges of only 35 miles and 73 miles respectively.[124]

This shortage of battery-stored energy affects many features that consumers take for granted. Running the air conditioner or heater draws energy from the battery, reducing driving range. Deliverable battery power is also a function of temperature. Electric car batteries perform poorly in cold weather, also reducing driving range.

Second, long refueling times are a major disadvantage, since large car battery packs are

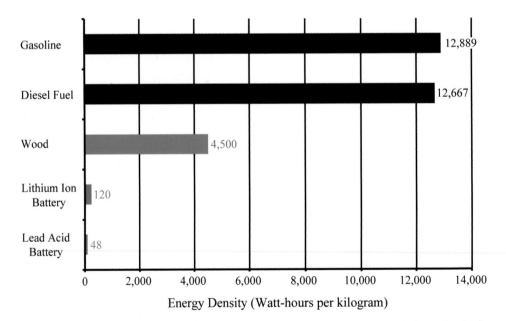

Energy Density of Batteries and Other Fuels. The energy density of today's batteries is tiny compared to gasoline, diesel fuel, and other fuels. (IOR Energy, 2009; UTC Center for Energy, 2011)[125,126]

slow to charge. Fuel pumps typically dispense gasoline or diesel fuel at the rate of seven to eight gallons per minute.[127] Gasoline fuel fill-ups require two to three minutes for small cars or four to five minutes for SUVs. In contrast, current charging times from a typical 120V home outlet approach twelve hours. Purchase of a 220V charger for $2,000 can reduce the time to about six hours for a fully depleted battery. Even commercial 440V charging stations will require a driver to leave the vehicle for 30 minutes or longer.[128]

Sticker shock will be a major hurdle for electric car buyers. GM's Chevy Volt was list priced in 2011 at $41,000 and the Nissan Leaf at just under $33,000.[129] After government rebates, these prices are still about $10,000 above competitive gasoline models. A 2009 study by the National Academy of Sciences estimated that a PHEV with a 40-mile range cost the manufacturer $14,000 to $18,000 more to build than a standard car of comparable size. The study projected that this cost differential would decline, but an electric would still cost $10,000 more to manufacture by the year 2030.[130]

ANOTHER *GREEN ENERGY* DEBACLE!

GM to Investigate After Chevy Volt Hybrid Catches Fire for SECOND Time in a Week—Even Though It Was Unplugged
—*Daily Mail*, Feb. 23, 2012[131]

But the biggest disadvantage of electric cars may be the poor lifetime of battery packs. Batteries are based on a chemical imbalance, a separation of charge that produces the electrical potential. Chemical reactions within the battery are always at work to remove the battery charge. The lithium ion battery in my laptop computer needed replacement after only two and a half years.

The day an electric car is driven out of the showroom, its lithium ion battery begins to degrade. The warmer the surrounding temperature, the faster the battery degrades. The more often the battery is recharged, the faster it degrades. The faster a battery is charged, the faster it degrades. If the battery is stored at full rather than partial charge, it also tends to degrade faster. That batteries degrade over time is as certain as death and taxes. In contrast, while the engine in a gasoline-powered vehicle ages, the fuel itself does not lose its power.

If you drive your electric car to work every day, and recharge it every other day, expect to need to replace the battery pack in three to five years. Senior Vice President Andy Palmer of Nissan Great Britain estimates the cost to

WISDOM FROM CLIMATISM?

...it ought to be possible to establish a coordinated global program to accomplish the strategic goal of completely eliminating the internal combustion engine over, say, a twenty-five-year period."
—Al Gore (1992)[132]

replace the Nissan Leaf battery pack at about £19,000 ($29,000).[133] The Chevy Volt has a smaller battery, but the cost will still be over $10,000. Auto manufacturers provide a 75,000 to 150,000 mile warranty on batteries, but since today's cars are usually driven for ten years or more, someone must still pay the cost for this replacement.

So is it reasonable to think that PHEVs can change car markets and reduce emissions in the near term? Not very likely. In 2011, despite marketing fanfare and government incentives, US sales of the Volt and Leaf were both under 10,000 units.[134] This is a drop in the ocean compared to the five-year average of more than 13 million US car sales per year. Even higher sales volumes will not have a noticeable effect on emissions, since the majority of the electricity used is produced in power plants burning coal or gas. Electric cars may be the future if superior batteries can be developed, but they are not the near-term future.

SUBSIDY MOUNTAINS: CLIMATE MADNESS IN EUROPE

For the last two decades, European nations promoted renewable energy with an almost missionary zeal. Subsidies, mandates, complex carbon and renewable certificate programs, and feed-in tariffs have been enacted by European governments. European leaders touted green renewable programs as not only a method to reduce emissions but also a way to produce economic growth and "green jobs." But as Sir Winston Churchill said, "However beautiful the strategy, you should occasionally look at the results."[135] Let's look at climate madness in Spain, Germany, and the United Kingdom.

Spain—A Model for the US?

Spain established feed-in tariffs in 1994 for wind and solar energy and then boosted these tariffs in 2004. Providers of wind-generated electricity were paid 50 percent to 97 percent above the average market price, and solar electricity generators were paid five to ten *times* the market price. Spanish utilities were required to buy the generated electricity, and subsidies were committed for 25 years.[136]

With guaranteed prices and markets, every Spanish hombre entered the market to get a piece of the subsidy pie. By 2008, the renewable building frenzy placed Spain globally in second place for installed solar capacity and third place for installed wind capacity. Spain was hailed by many as the model to follow, including President Barack Obama, who stated: "Spain generates almost 30 percent of its power by harnessing the wind."[137]

But what did this renewable build accomplish? A study by Dr. Gabriel Calzada Álverez, an economics professor at the University of Rey Juan Carlos, characterizes the effort as a "renewables bubble." The renewables program was largely responsible for a 61-percent increase in electricity rates from 2000 to 2008. At the same time, the government accumulated a renewable energy subsidy obligation of €28.7 billion ($36 billion). Dr. Calzada's study found that the renewables program actually *reduced* the number of jobs in the Spanish economy. According to his estimates, about 50,000 jobs

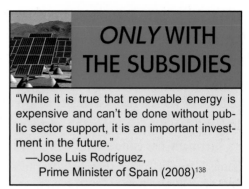

ONLY WITH THE SUBSIDIES

"While it is true that renewable energy is expensive and can't be done without public sector support, it is an important investment in the future."
—Jose Luis Rodríguez,
Prime Minister of Spain (2008)[138]

were created, but at the horrendous cost of €571,000 ($723,000) per job. At the same time, the program increased electricity rates and accumulated government liabilities, costing the whole of the economy 113,000 jobs that would have been created elsewhere. For every green job created, an estimated 2.2 jobs were lost in other industries.[139] During this renewable build, Spain's output of greenhouse gas emissions *increased* 53 percent from 1990 to 2006, the highest rate of increase in Europe.[140]

The Spanish "renewables bubble" burst in 2008. Unable to cope with the mounting subsidy obligations, the government announced a retroactive 30-percent cut in subsidy payments and established a quota process for solar systems. Market demand for solar cells dropped 80 percent in a single year.[141] Over 20,000 green jobs disappeared. In early 2012, the government announced a suspension of all subsidies for wind, solar, and renewable energy, resulting in additional job losses.[142]

Do Renewables Create Job Growth?

A story about Dr. Milton Friedman, Nobel laureate in economics, sheds light on the question of renewables and jobs growth. As the story goes, Dr. Friedman was visiting a canal construction site in an Asian nation. He saw that all of the workers were using shovels and that there was a lack of heavy construction equipment. He asked a government official who was with him, "Why don't the workers have heavy machinery?" The official replied, "You don't understand. This is a jobs program." Friedman reportedly said, "Well, then you should take away their shovels and give them spoons."[143]

The point is that economic growth, not jobs, should be the goal of governments. Economic growth is created by increases in productivity—that is, higher output of goods and services from each worker. Since renewable energy is more expensive and intermittent, forced adoption of renewables reduces productivity in the energy sector. Other industrial sectors that rely on low-cost energy incur higher costs, also reducing the productivity of these sectors. The net result of forced adoption of renewable energy is a loss in productivity throughout the economy, a reduction of economic growth, and a net loss of jobs.

Germany—Welcome to Environmental Madness

Germany is the renewable energy leader of Europe. In 2010, Germany had over 21,000 wind turbine towers in operation, good for third place in the world behind China and the United States.[144] The nation's biofuel usage is the highest in Europe, providing about six percent of vehicle fuel. Germany is also the largest producer of biomass-fired electricity in Europe, doubling Sweden, the next largest producer.[145]

Germany has gone solar. From a solar cafe in Kirchzarten, to a solar golf course in Bad Saulgau, to the largest solar boat in the world, the Alster Sun of Hamburg,[146] blind faith in solar energy has characterized German policies for the last 30 years. More than 1.1 million solar systems have been installed,[147] a total greater than any other nation.

Generous government support for renewables began in 1991 with the Electricity Feed-In Law. A follow-on law in year 2000 guaranteed price support for 20 years. Utilities are required to buy energy from renewable producers at rates far above the market rates of two to seven Eurocents per kW-hr. Onshore wind turbines were granted tariffs that were double the market rate, and solar energy producers received a tariff of over 40 Eurocents per kW-hr. ($0.51 per kW-hr.), about eight times the market rate.[148]

Over the last 20 years, Germany has built a sizable renewables industry. Wind and solar energy systems have become major export businesses. According to the Ministry of the Environment, Nature Conservation, and Nuclear Safety, by 2010 the renewables sector included 367,000 jobs,[149] but as we will discuss, the cost of these jobs is high.

In the fall of 2010, the nation adopted the policy of Energiewende (Energy Transition), a plan to move from an economy based on hydrocarbon and nuclear fuels to one based on renewable energy. The plan calls for renewables to achieve a 30-percent share of energy usage by 2030 and a 60-percent share by 2050. The plan further calls for energy consumption to fall by 20 percent by 2020 and 50 percent by 2050, compared to 2008.[150]

WISDOM FROM CLIMATISM?

"The transformation towards a low-carbon society is therefore as much an ethical imperative as the abolition of slavery and the condemnation of child labor."
—German Advisory Council on Global Change (2011)[151]

Along with Britain, Germany is the European co-champion for the ideology of Climatism. The thinking behind this policy is captured in *World in Transition: A Social Contract for Sustainability*. This report was produced in March 2011 by the German Advisory Council on Global Change, a group that is a close advisor to the government of Chancellor Angela Merkel. The report warns that the carbon-based economic model is "an unsustainable situation" and states that "decarbonization of energy systems on a global level is feasible." It calls for global carbon pricing, mandatory emissions restrictions on nations, and a "global governance architecture" to force compliance. The study also calls for the phase out of nuclear power along with a massive emphasis on renewable energy.[152]

On March 11, 2011, an earthquake measuring 9.0 on the Richter magnitude scale struck just off the northeast coast of Japan. Fifty minutes later, a tsunami hammered the six nuclear plants in Fukushima, Japan, eventually causing cooling system failures, explosions, and severe damage at the three plants that were in operation at the time. It took almost two months for the government of Japan to gain control of the situation. A zone of 12 miles (20 kilometers) around the site was designated as a restricted area, forcing the evacuation of 100,000 people.[153]

In the wake of the Fukushima incident, in May 2011 the Merkel government announced a complete phase out of nuclear power by the year 2022.[154] By July, eight of Germany's seventeen reactors had been shut down. Over a period of just three months, Germany went from a net exporter to a net importer of electricity.[155]

Can Germany make a transition to renewables, shut down nuclear, and maintain economic prosperity at the same time? A rational observer should think it unlikely. In 2010, Germany received 89 percent of its energy from hydrocarbons and nuclear power. Oil provided 33 percent, coal provided 23 percent, natural gas provided 22 percent, and nuclear provided 11 percent of energy usage. After 20 years of subsidy, mandates, and promotion, renewables provided only 11 percent of the nation's energy.[156]

Cracks are starting to appear in the renewables facade. Despite more than a million installations, solar provided *less than one percent* of Germany's energy supply in 2010. Germany is located in northern Europe. Kassel, a city in central Germany, sits at 51°N latitude, about as far north as Calgary, Canada. The German sun field is at a low angle and often interrupted by overcast skies. In 2010, total electricity output from Germany's photovoltaic solar systems was only 7.7 percent of rated output! Also in 2010, wind turbine towers provided only 1.5 percent of Germany's total energy needs, operating at a meagre 16-percent capacity factor.[157] With wind and solar systems operating less than one-fifth of the time, how will hydrocarbon and nuclear plants be replaced?

Despite these small and intermittent outputs, Germany continues to spend vast amounts on the mirage of renewables. When a solar system is installed, utilities are committed to pay the solar feed-in tariff for 20 years, a cost that is passed on to electricity customers. In 2010, solar providers collected more than €8 billion ($10 billion) in subsidy payments. By the end of 2010, committed payments to solar providers exceeded €100 billion ($127 billion).[158] Manuel Frondel, a researcher at RWI Institute, estimates that every job in the solar sector costs consumers €250,000 ($316,000).[159]

It appears that the German renewable energy train is headed for a crash. In 2010, Berlin citizens paid about 23 eurocents per kW-hr for electricity, the second highest rate in Europe, and the rate is rising.[160] A survey of 1,520 companies in 2011 by the German Chamber of Industry and Commerce found that:

> One fifth of every industrial company has moved activities to foreign countries, or plans to do so, because of the uncertain energy and raw material supply.[161]

The government now realizes that it can't continue to mandate high feed-in tariff rates for new solar systems. Economic Minister Philip Roesler admitted that spiraling costs from solar subsidies are a threat to the economy, stating that "the increase in installations in the past few years has gone far beyond what we had targeted." As a result, the government cut feed-in tariff rates three times in 2011, with more cuts planned for 2012 and plans for complete phase-out of subsidies by 2017.[162]

The subsidy cuts could not have come at a worse time for German solar cell manufacturers. Due to oversupply and subsidy cuts across Europe, the price of solar cells fell more than 50 percent from 2011 to 2012. In 2004, Germany held about 70 percent share of the global solar panel business, but this share declined to about 20 percent in 2011, mostly lost to Asian suppliers.[163] But even Asian low-cost suppliers reported that they could not make a profit at 2011 solar cell prices of $0.50 per watt.[164] Dozens of solar cell companies went bankrupt, including Solon and Solar Millennium, once considered model German companies. After all the hoopla, it appears that 1 percent of Germany's energy is about the limit for solar.

It seems like the "will of the wisp" wind and the trivial output of solar are poor foundations for the future German economy. But Chancellor Merkel is undeterred. She states:

> I think we can say our energy system will be the most efficient and environmentally friendly in the world.[165]

United Kingdom—Renewable Folly of Her Majesty's Government

Speaking to his staff in 2010, Prime Minster David Cameron stated a goal: "I want this to be the greenest government ever."[166] The folly of pursuing renewable energy to stop global warming reigns supreme in the UK. On October 28, 2008, the House of Commons passed the Climate Change Act, calling for an 80 percent cut in emissions by the year 2050 from the 1990 level. The act passed by a vote of 463 to 3, even though London's first October snowfall since 1922 was occurring right outside the windows of Westminster Palace.[167]

THE GORE EFFECT

Snow Blankets London for Global Warming Debate
"Snow fell as the House of Commons debated Global Warming yesterday—the first October fall in the metropolis since 1922."
—*The Register*, Oct. 29, 2008[168]

The United Kingdom has huge renewable goals, but little renewable output so far. According to the Department of Energy and Climate Change (DECC), in 2010 the UK got 97 percent of its energy from hydrocarbons and nuclear. Of that energy, 43 percent came from natural gas, 32 percent from oil, 15 percent from coal, and 7 percent from nuclear. Only 3 percent came from renewable sources.[169] The nation has a goal of 15 percent of energy and 31 percent of electricity from renewables by year 2020.[170]

To try to reach the goal, high levels of subsidy and mandates are used to promote renewables. The Renewables Obligation (RO) is the primary tool. The RO is a complicated system requiring electrical utilities to buy an increasing share of energy from renewable sources. Renewable sources such as wind, solar, and biomass are issued RO certificates. Utilities must buy certificates from renewable companies and

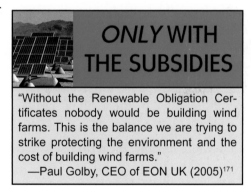

ONLY WITH THE SUBSIDIES

"Without the Renewable Obligation Certificates nobody would be building wind farms. This is the balance we are trying to strike protecting the environment and the cost of building wind farms."
—Paul Golby, CEO of EON UK (2005)[171]

deliver these certificates to the government or make a compensating payment. The system effectively raises the market price for electricity and provides subsidies to renewable sources.

The United Kingdom is engaged in a massive, heavily subsidized wind turbine-building program. From 2007 to 2010, national wind turbine capacity more than doubled.[172] The renewables subsidy, which is overwhelmingly for wind and is paid from consumer electricity bills, equaled £1.1 billion ($1.7 billion) in 2010.[173] The government wants to spend a staggering £100 billion ($154 billion) by 2020 to reach its goals for renewables. In a report by the Centre for Policy Studies, analyst Tony Lodge pointed out that the 2020 renewables goal would cost the average UK household £4,000 ($6,153).[174] When British citizens realize that mankind isn't destroying the climate, think they'll go for this plan?

A massive wealth transfer is underway in the UK, courtesy of the wind program. Taxes and higher electricity bills are paid by British common citizens. The government receives taxes and RO payments from utilities and gives subsidies to wind energy companies. Wind companies then pay high rents to British nobility, who own the land on which the wind turbines are built. Annual payments to British dukes, earls, and duchesses amount to £20,000 ($31,000) per year *for each wind turbine.*[175] It's easy to see why the British landed class approves of green energy.

And yes, Her Majesty's Government has also been subsidizing the installation of solar systems in the United Kingdom. Feed-in tariffs were established that paid a princely twelve times the market rate for electricity. DECC data now shows that solar systems in the UK produced only about *5 percent* of rated output in 2010, but who cares?[176] A nation *must* have solar to be truly green. At an average 53°N latitude, solar in England is as useless as mango farming in Minnesota.

Indeed, when the government decided to cut solar subsidies in 2011, there was a loud outcry. Developers claimed that with the reduced subsidy, it would take homeowners "25 years to earn back their investments."[177] Where is Sir Winston when we need him?

THE MYTH OF PEAK OIL

Environmentalists have long argued that we are running out of hydrocarbons. This concern is captured by the myth of "peak oil." The concept of peak oil was originally defined by US geologist Marion King Hubbert to describe a bell-shaped curve for US oil production. More broadly, the term refers to the idea that world oil resources (and hydrocarbon resources) are finite, and it raises fears that we soon will run out of oil and other hydrocarbon fuels.

The forces of Climatism have seized upon peak oil as another reason why society must adopt renewable energy. The argument is that, since Earth will soon run out of oil and gas, nations must mandate the adoption of renewables. Further, governments must actively force the adoption of renewables.

It's true that eventually global production of oil and other fuels must peak, but why not let economics govern such a situation? As oil and gas get scarce, prices will rise, other energy alternatives will become relatively more cost effective, and consumers and businesses will adopt other alternatives.

But peak oil alarmists (who are hard to tell from Climatists) don't see it that way. Somehow, passing the point of peak oil will cause a global cataclysm. David Cohen of the

---

A FEW OF MANY FAILED PREDICTIONS ABOUT PEAK OIL AND GAS

"...the amazing exhibition of oil which has characterized the last twenty years, and will probably characterize the next ten or twenty years, is nevertheless, not only geologically but historically, a temporary and vanishing phenomenon..." —J.P. Lesley, State Geologist of Pennsylvania (1886)[178]

"It is unsafe to rest in the assurance that plenty of petroleum will be found in the future merely because it has been in the past." —American Association of Petroleum Geologists (1936)[179]

"Natural gas is an ideal fuel, but one that is in short supply...If consumption of natural gas continues at the present rate and exploration does not pick up, the United States may burn its last molecule of domestic natural gas within 20 years." —Author Allen Hammond (1973)[180]

"Prominent exploration experts have recently predicted that total world production of liquid oil will peak by about the end of this decade—or a few years later if production does not rise much—and will decline thereafter." —Environmentalist Amory Lovins (1975)[181]

"Oil production should peak out around the world in the early 1990s...That means in five years' time we may have chewed up most of the possibility of further expansion of oil production." —Former US Energy Secretary James Schlesinger (1977)[182]

"Gas lines and rapid increases in oil prices during the first half of 1979 are but symptoms of the underlying oil supply problem—that is, the world can no longer count on increases in oil production to meet its energy needs." —US Central Intelligence Agency (1979)[183]

Association for the Study of Peak Oil and Gas (ASPO) warned in 2007:

> "Although it is impossible to predict the future, extrapolating present trends...leads to "The Perfect Storm" sometime in the next decade. At the tipping point, oil prices exceed the pain tolerance of a sufficient number of global consumers, causing economies to roll over into severe recession."[184]

The ASPO also supports the theory of man-made global warming.

A number of organizations, including the International Energy Agency, claim that world crude oil production peaked in 2006. The DOE disagrees, showing global production at 86.8 million barrels per day in 2010, up 2.5 percent since 2006. More interesting are trends in global oil reserves and consumption over the last 30 years. In 1980, the world was consuming about 63 million barrels per day, with estimated oil reserves of 642 billion barrels, or about 28 years of supply at the 1980 usage rate. Over the next 28 years, both usage and proven reserves increased. By 2009, world crude oil consumption had increased by 34

"Civilization as we know it will come to an end sometime in this century unless we can find a way to live without fossil fuels."
—Professor David Goodstein (2004)[185]

percent to 85 million barrels per day. But world oil reserves had increased even faster. World oil reserves in 2009 totalled over 1.3 trillion barrels, a 43-year supply at the higher usage rates (opposite page).[186]

World gas reserves showed a similar powerful increase. In 1980, the world had natural gas reserves to cover 49 years of supply at 1980 rates of consumption. Even though gas consumption increased by 97 percent from 1980 to 2009, world gas reserves increased faster, amounting to 60 years of supply in 2009 at the higher levels of demand.[187] The trend curves show no indication that world production of either oil or gas is in serious trouble.

It's true that bubble-up-in-your-backyard crude is no longer available. Most of the easily accessible sources of petroleum have been exhausted. But it's also true that man's ingenuity, driven by $100-per-barrel prices, can access many oil sources that were deemed worthless only decades ago. Today, crude oil is pumped from wells above the Arctic Circle, pumped from the ocean floor at depths greater than 5,000 feet, and mined from the oil sands of Canada. And the hydraulic fracturing revolution has just begun.

THE HYDRAULIC FRACTURING REVOLUTION

In October 2007, six members of Greenpeace scaled the 200-meter-high tower of the Kingsnorth power station in Kent, UK, to protest government plans to build a new coal-fired plant at Kingsnorth.[188] The Sierra Club claims to have stopped the construction of more than 100 coal-fired plants in the United States.[189] Fall of 2011 brought protests in front of the White House over the proposed Keystone pipeline project, which would have delivered crude oil from Canadian oil sands to the US. Dr. James Hansen was arrested as part of the protests, and President Obama subsequently decided to delay approval of the pipeline.[190] These are just a few examples of efforts by the forces of Climatism to halt the use of hydrocarbons around the world.

But all these efforts to save the planet are now threatened by a new revolution in the production of gas and oil: the hydraulic fracturing revolution. Hydraulic fracturing, or

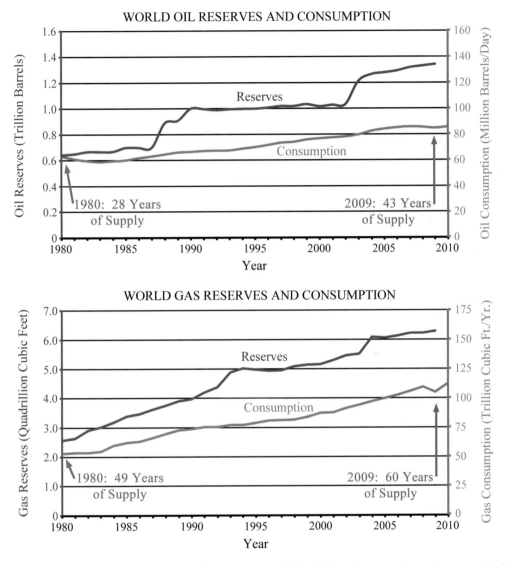

World Oil and Gas Reserves and Consumption 1980–2010. The top chart shows world oil proven reserves (left axis) and daily consumption (right axis). The bottom shows world gas proven reserves (left axis) and annual consumption (right axis). (US Department of Energy, 2012)[191]

"fracking," is a decades-old technique for unlocking hydrocarbons from shale rock deep below the surface of Earth. New innovations in fracking have the potential to deliver almost limitless quantities of gas and oil to people at affordable prices.

Fracking technology was recently refined by Texas energy baron George Mitchell. For more than ten years he worked to find a way to access natural gas locked in the Barnett

Shale, a shale field located more than 5,000 feet below the city of Fort Worth and fifteen counties in north-central Texas. Mitchell spent millions of dollars and ignored others who said shale gas would never be profitable. In 1997, Mitchell's team found the key. Water, sand, and a small amount of chemicals, injected under pressure, could fracture the shale and create millions of tiny fissures, releasing the trapped gas. To develop a large producing field, Mitchell used horizontal drilling to bore mile-long horizontal shafts into the shale.[192]

The results have been spectacular. In year 2000, shale gas was only one percent of US gas supplies; a decade later it had grown to 25 percent. Prior to fracking, US gas supplies were dwindling, prices reached $15 per million British thermal units (Btu), and port facilities were being constructed to import liquified natural gas. By 2011, gas prices had fallen to $4 per million Btu. Port facilities were being refitted for anticipated gas exports.[193]

Volume production of shale gas is now underway from the Barnett Shale field in Texas, the Haynesville Shale field in Lousiana, the huge Marcellus Shale field under the states of New York, Pennsylvania, West Virginia, and Ohio, and other US sites. In Pennsylvania alone, more than 2,000 fracking wells were drilled from 2008 to 2011.[194] Hundreds of thousands of jobs have been created nationally—none requiring a government subsidy.

Natural gas is recovered from pores in the rock matrix. Prior to the hydrofracturing revolution, most gas came from coarse rock formations, such as sandstone. Shale, a "tight" formation, is now a source for natural gas. Enormous shale fields cover much of Earth's surface. Potentially large shale gas fields exist in Argentina and other nations of South America, Australia, Canada, several nations of Europe, Mexico, Russia, South Africa and other nations of Africa, and the United States. A 2010 report from the DOE states that "the international shale gas resource is vast." The report goes on to estimate that hydraulic fracturing more than doubles the world reserves of gas to over 200 years of technically recoverable reserves.[195]

Environmental groups have recently changed their tune about natural gas. Once hailed as an interim solution because combustion of gas emits lower emissions of CO_2 than coal, the huge potential of fracking is now viewed as a grave threat to renewable energy programs.

In 2011, energy exploration company Cuadrilla Resources announced a natural gas find of an estimated 200 trillion cubic feet near Blackpool in northwest England.[197] The find

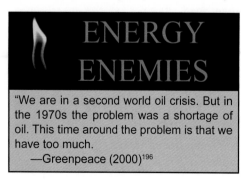

ENERGY ENEMIES

"We are in a second world oil crisis. But in the 1970s the problem was a shortage of oil. This time around the problem is that we have too much.
—Greenpeace (2000)[196]

has the potential to restore the UK to self-sufficiency in natural gas and eliminate a growing dependency on imports.

The negative outpouring of press regarding the Cuadrilla announcement was remarkable. News articles warned of earthquakes, tap water you could set on fire, radioactive water pollution, and runaway climate change. A 2011 report from the Tyndall Centre for Climate Research opposed the pursuit of shale gas, warning:

> ...shale gas offers no meaningful potential as even a transition fuel. Moreover, any significant and early development of the industry is likely to prove either economically unwise or risk jeopardising the UK's international reputation on climate change.[198]

The 2010 documentary *Gasland* painted a picture of water contamination, environmental damage, and health dangers from hydraulic fracturing. It won numerous documentary film awards and was embraced by much of the environmental community. One movie scene showed gas from a water tap in Weld County, Colorado, igniting and exploding into flame.[199]

But *Gasland's* portrayal of hydraulic fracturing was not accurate. Fracturing occurs at a depth of 5,000 to 8,000 feet, separated from the water table by thousands of feet of rock, making water contamination highly improbable. Proper techniques must be followed to prevent gas from leaking from vertical shafts, but these procedures are routine in the industry. More than 500,000 wells have been drilled using fracturing in the US over the last 30 years, and the documented number of water pollution incidents can be counted on one hand, with most fingers remaining. The flaming Colorado tap from *Gasland* appears to be from natural methane mixing with water. A 1976 report by the Colorado Division of Water found problems with methane in Weld County water, long before any fracking in the area.[200]

A battle over hydraulic fracturing is spreading across the world. Local communities in New York state have banned fracking. The parliament of France banned fracking in 2011. Poland and China are pushing forward full speed to access shale gas. Debate rages in Canada, Germany, the UK, and other nations.

It's a battle between common sense and ideology. The facts show that shale gas is a low-cost, high-energy-density fuel that can be recovered with minimal impact to the environment, providing centuries of power for mankind. But Climatist ideology continues to push for poor wind, solar, and biofuel alternatives that require government subsidies and mandates. Thomas Jefferson was correct when he said:

> It is error alone which needs the support of government. Truth can stand by itself.[201]

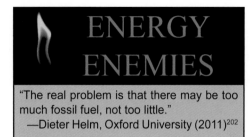

Hydraulic fracturing techniques are now also being used for the recovery of "tight oil." Tight oil is petroleum contained in shale fields that can be accessed using horizontal drilling and fracturing techniques. Since 2006, fracking has been used to make the Bakken Shale oil field in North Dakota the fastest-growing field in the United States. US oil production rose in 2009 and 2010 for the first time in 20 years due to output from shale fields.[203] Shale oil is now being pursued in Australia, Canada, China, France, India, and Poland. As energy expert Robert Bradley says:

> We are not running out of oil and gas. We are running into oil and gas. Predictions of peak oil and gas are not only being made for the wrong year and decade, but also for the wrong century.[204]

THE RENEWABLE ENERGY MIRAGE

We would all like to believe that renewable energy is the answer, but it isn't. The renewable energy revolution proposed by Climatism is a mirage. The world remains overwhelmingly based on hydrocarbon fuels. The huge volume of global energy usage means that dilute, intermittent, and costly renewables will be unable to provide a significant share of global energy usage for decades to come, if ever.

But the good news is that mankind is not running out of hydrocarbon energy sources in the short term. On the contrary, the hydraulic fracturing revolution holds the promise of centuries of supply of natural gas and possibly oil, if the world can shake the misguided grip of climate madness.

CHAPTER 12

YOU CAN'T MAKE THIS STUFF UP!

"Sometimes the very learned and clever can be brilliantly foolish,
especially when seized by an apparently good cause."
—CARDINAL GEORGE PELL OF AUSTRALIA

Climatism is accepted by today's society. This results in all sorts of foolish behavior. Although somewhat hard to believe, all of the topics in this chapter are true examples of climate madness. But these are just a few of countless examples from our mad, mad, mad world.

POUR WHAT'S LEFT OF GRANNY ON THE ROSES

Cremation is a serious climate problem. More than 10 million people are cremated each year, and this number is rising.[1] Between the energy used for the heating of cremation furnaces to 850°C and the emissions from the burned body and coffin, up to 160 kilograms (350 pounds) of emissions are produced per corpse.[2] Worldwide, this amounts to more than 1.6 million tons of CO_2 emissions each year. Imagine how many polar bears are dying because of this environmentally harsh process.

But there is a climate-safe solution to the problem. Aquamation Industries, a company in Australia, has developed an aqualine hydrolysis process that chemically dissolves a body in four hours. The firm claims: "Aquamation is a more natural, ethical, and environmentally friendly alternative to cremations." After the process, the resulting solution is "safe enough to pour on the rose bushes." Oh, concerned about your pet? The firm also sells small aquamation units to eco-dispose of our "departed loved pets."[4]

ANOTHER HIGH-POTENTIAL ENERGY SOURCE?

Crematorium to Heat UK Swimming Pool
"A town near London plans to warm its new sport center and public pool with heat from a nearby crematorium...'This form of energy is certainly renewable, unless the locals stop dying'...'The use of waste heat energy in this way is good practice and very innovative. It would genuinely be a first in the UK and demonstrates Redditch Borough Council's seriousness about addressing climate change issues.'"
—*Grist*, February 4, 2011[3]

BREAKTHROUGH: SPAIN PRODUCES SOLAR ENERGY AT NIGHT!

A letter from Pedro Marin, Spain's Secretary of State for Energy, revealed that Spanish solar power providers had done the impossible. From November 2009 to January 2010, solar installations delivered 4,500 megawatt-hours of electricity to the grid between midnight and 7 a.m. Environmental groups rejoiced at the prospect of global application of this technology to solve the world's energy problems.

Alas, the Spanish solar-panel trade group El Mundo determined that solar stations may have been run on diesel fuel generators and the electrical output sold as solar energy. This innovation allowed

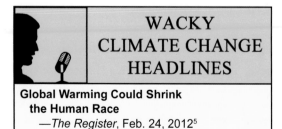

WACKY CLIMATE CHANGE HEADLINES

Global Warming Could Shrink the Human Race
—*The Register*, Feb. 24, 2012[5]

operators to receive solar tariff rates many times the market rate for electricity. It's likely these entrepreneurs were also running the generators during daylight hours to boost revenues from "solar energy."[6]

GREEN, LEAFY FIGHTER JETS THAT FLY ON ALGAE

The United States military, the world's most powerful military force, has become a tool of Climatist ideology. US Secretary of the Navy Ray Mabus said:

> We're gonna be using American produced, American energy that will create jobs in the United States, will create a far more secure source of energy for us and will make us better environmental stewards because we will be contributing less to climate change and burning much cleaner fuel.[7]

The Department of Defense (DoD) is the largest consumer of petroleum-based products in the United States, using about 340,000 barrels per day of petroleum-based fuels, which is 93 percent of US national government usage. About 95 percent of DoD fuel usage is distillates, fuels which differ from commercial gasoline and diesel fuels. Military distillate specifications require higher ignition "flash points" than commercial fuels, providing improved safety for military operations. Aircraft jet fuel and fuel for naval vessels are the two largest applications for military distillates.[8]

The US Air Force and Navy have established goals to use 50/50 blends of biofuel and petroleum-based fuel for planes and ships. The Air Force plans to obtain one-half of its fuel from alternative blends by 2016. The Navy's goal calls for a "Green Strike Group" task force fueled by the 50/50 blend by 2016 and for alternative fuels to power half of all energy consumption by 2020.[9]

Military fuels today are produced by distillation processes at petroleum refineries. JP-5 jet fuel used by the Navy costs about $4 per gallon, a little more than commercial gasoline.[10] Alternative substitutes that meet military specifications can be produced from coal, natural gas, or biomass using a Fischer-Tropisch distillation process.[11] But these alternatives will cost taxpayers a wee bit more than petroleum fuels.

In December 2012, the Navy placed the "largest single purchase of advanced drop-in biofuel in government history," a $12-million order to three firms to deliver a fuel derived from algae for the Green Strike Group. Algae-based biofuel will be purchased at $26 per gallon and 50/50 blended with petroleum-based fuel to provide a green fuel at the bargain price of $15 per gallon, almost four times the price of current fuel.[12] To support development of the algae-to-biofuels process, the Navy and the Departments of Energy and Agriculture announced that they would invest $510 million to create an "advanced biofuels industry." Algae is now the leading candidate feedstock.[13]

Why try to run US planes and ships on fuel from algae? It's not for better performance. A 2011 study by the Rand Corporation for the DoD concluded that:

> "...the use of alternative, rather than petroleum derived, fuels offers no direct military benefits."[14]

Using algae as feedstock for the Fischer-Tropisch process is expensive and unproven, compared to coal feedstock, which can deliver military distillates at reasonable prices. If the goal is to reduce military dependence on Middle East oil, this could be done by producing fuel from US coal at much lower cost.

The reason is greenhouse gas emissions, of course. The military has swallowed the ideology of man-made warming and is willing to invest millions on unproven processes and pay enormous per-gallon prices just to try to reduce those evil greenhouse gases. And US taxpayers will pay for this green miracle.

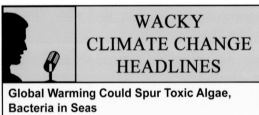

WACKY CLIMATE CHANGE HEADLINES

Global Warming Could Spur Toxic Algae, Bacteria in Seas
—*Wild Singapore News*, Feb. 20, 2011[15]

Scientists recently warned that man-made global warming could spur the growth of toxic algae in the oceans. Can toxic algae be used for biofuel?

STOP SIGNS CAUSE GLOBAL WARMING?

The city of Scarborough-Agincourt, a suburb of Toronto, Canada, recently considered a gripping climate question. Were additional stop signs proposed for traffic control contributing to global warming? The concern was that additional stop signs would force vehicles to stop more often, releasing increased greenhouse gases, a threat to the planet. The City of Toronto considered this question for two months but was unable to come to a conclusion.

The Canadians may be on to something here. Imagine how many tons of additional CO_2 is emitted by vehicles reacting to poor traffic planning across the world. As developing nations add more vehicles to their cities, this could be a growing emissions problem.

Maybe the United Nations should form an organization to advise developing nations on "traffic planning to minimize greenhouse gas emissions."[16]

THE PROBLEM OF AUSTRALIAN CAMEL GAS

More than 1.2 million wild camels roam the outback of Australia. Called feral camels, these nomads munch up the foliage and then expel about 45 kilograms of methane per camel each year, from either the nose end or tail end. According to the IPCC, methane is a more potent greenhouse gas than carbon dioxide, so each camel produces about one ton of CO_2-equivalent emissions per year.[17] This makes the problem of Australian camel gas almost as severe as emissions from worldwide human cremation.

The Carbon Farming Initiative Act (CFI) became law in Australia on December 8, 2011. Among many measures in the law to fight climate change, the CFI calls for:

> The reduction of methane emissions through the management, in a humane manner, of feral goats, feral deer, feral pigs, or feral camels.[18]

By "management," they mean shooting the poor beasts.

The innovative firm Northwest Carbon lobbied for the camel culling clauses in the CFI, working closely with the Australian Department of Climate Change and Energy Efficiency. Now that the culling of feral animals is approved by the government, Northwest Carbon will be "humanely" shooting goats, deer, pigs, and camels from helicopters.

Australia is a pace-setter in this new process for reducing global emissions. Imagine the emissions savings we can amass if we extend "culling technologies" to elephants, giraffes, gazelles, buffalo, elks, and many other species. Why, we probably could even get the World Wildlife Fund to agree to stepped-up hunting of polar bears, all in the name of fighting climate change.

CLIMATIST HIT LIST

The Australian Feral Camel[19]

ECO-FRIENDLY BIRD-CHOPPING WIND TURBINES

Wind turbines kill birds. Not just a few sparrows, but large quantities of birds, including great numbers of hawks, owls, and eagles. A study for the California Energy Commission concluded that roughly 1,000 raptors were being slaughtered each year at the Altamont Pass wind farm in California.[20] In 2009, the US Fish and Wildlife Service estimated 440,000 bird deaths from US wind farms, about 20 birds for every turbine operating at the time.[21] Extending this estimate worldwide places 2010 global bird deaths from wind turbines at over three million per year.

Eagle killed by a wind turbine in Sweden.
Photo by Hedgren[22]

In 2008, the Southwell Community Primary School in Portland, England, erected a wind turbine 9 meters (30 feet) high to generate green electricity for the school. The school received a grant from the Department of Energy and Climate Change to allow it to purchase the £20,000 turbine tower. After installation, the turbine generated an average six kilowatts of electricity, about 40 percent of the school's requirement.[23]

But then disaster struck. The turbine began killing seagulls, slicing up a total of fourteen birds over a six-month period. Headteacher Stuart McLeod arrived early each morning to remove dead birds before the children arrived. The school tried everything to solve the problem. They proposed to paint the turbine blades "dazzle yellow," but the manufacturer said no. They investigated bird-scaring plastic owls and consulted seabird experts, all to no avail. Headteacher McLeod lamented, "We've tried so hard to be eco-friendly but now we can't turn it on."[24]

WACKY GREEN EDUCATION

Like the Southwell Community Primary School, other schools are striving to be pioneers for a green planet. In September 2004, former UK Prime Minister Tony Blair stated:

All new schools…should be models for sustainable development: showing every child in the classroom and the playground how smart building and energy use can help tackle global warming…Sustainable development will not just be a subject in the classroom: it will be in its bricks and mortar and the way the school uses and even generates its own power.[25]

A zero-carbon Living Ark® classroom. It was too cold in winter to be used by students.[28]

Following Prime Minister Blair's vision, the Muswell Hill Primary School in North London purchased a "zero-carbon" Living Ark® classroom in 2010. The classroom was heralded as the way of the future, built at a cost of £25,000 ($38,000). The classroom was constructed from sustainable wood, sheep's wool, and soil. It included a mud and grass roof and was powered by solar cells.[26] The classroom was built as part of the Muswell Hill "Low Carbon Zone," one of ten such zones touted by the Mayor of London.[27]

But the Muswell school soon got a lesson about London sunshine. The solar cells only provided enough energy for light bulbs, and the classroom was bitterly cold in winter months. Liberal Democratic Councillor Gail Engerts was disappointed:

It is such a shame that, considering the fanfare, it emerges that this facility cannot be used by the children all year round.[29]

Maybe that's because you're in London and not Jamaica?

A school district in Boca Raton, Florida, sought to reduce water use and "diminish greenhouse gases" by installing waterless urinals in boy's bathrooms. The urinals used no water, trapped urine odors with disposable cartridges, and required only regular cleaning and changing of cartridges every three months. The district intended to save $100 in water costs per urinal each year, while helping the climate.

IS THIS WHAT YOU MEAN BY SUSTAINABLE???

Southern California wind turbine farm. Photo by McCauley (2005)[31]

But the green solution backfired at Spanish River High School. Without water to wash the urine into the sewer system, corrosive gases from the waste ate through the metal of the copper pipes. Leaking urine dripped into walls and onto floors. School board chairman Frank Barbieri remarked:

> It was pretty disgusting. The girls had to step over a river of urine. I could smell it as soon as I walked into the hallway.

The school district was faced with a $500,000 bill to clean up the damage at Spanish River High and to replace the waterless urinals with traditional urinals in four schools.[30]

CALIFORNIA: POURING MONEY DOWN A GREEN DRAIN

The State of California has long been a shining green example of sustainability on the US west coast. One California eco-leader is the Los Angeles Community College (LACC) system. In 2002, the Board of Trustees of LACC began a sustainability policy that grew into a $6-billion green-building program, including the construction of 87 new buildings on nine campus locations that met Leadership in Energy and Environmental Design Standards (LEED™).[32] While building to LEED standards may be good, the renewable energy plan of LACC was comical.

The facilities director of LACC proposed that the college spend almost a billion dollars in renewable projects that would allow LACC to disconnect from the electrical grid, even though the system spent less that $8 million on annual power bills. Wind turbines, solar panel systems, geothermal plants, and even hydrogen fuel cell systems were all proposed. Millions were spent on consultants before it was determined that many of the projects were not feasible.[33]

An example was the project for geothermal systems that was proposed for all nine sites.

Northwest Parking Lot solar farm at East Los Angeles Community College[35]

It was found that the geothermal system at the Pierce College site would save $15,000 annually at a cost of $1.3–2.4 million, requiring about 100 years to pay back invested capital. The LACC-appointed energy oversight committee deemed this cost prohibitive "unless substantial grants and incentives are available."[34] So, if the government subsidizes this project *then* a 100-year payback makes sense?

LACC scaled back most proposals, but did purchase several solar photovoltaic systems. One such system, the Northwest Parking Lot Photovoltaic Farm at East

Los Angeles Community College, cost $10 million to produce about one megawatt of nameplate power.[36] This electricity price is five times the cost of a commercial wind turbine farm on a per-megawatt basis.

In all, LACC spent $33 million on solar systems to reduce electricity bills by 7 percent, or about $600,000 per year.[37] If government subsidies are included, it would take over 70 years for the college to break even on the money spent. Since the solar panels will need to be replaced after 25 years, this project will *never* break even.

REALLY FUTILE CLIMATE GESTURES

On a local scale, nothing can top the principal at Ansford Academy in Castle Cary, England, for a really futile climate gesture. In order to teach students how to lower their carbon footprint, Principal Robert Benzie turned off the school heat during a 1°C (34°F) day in December 2011. Students and teachers shivered in near-freezing cold and tried to function while wearing winter coats and gloves. One teacher remarked, "I've never worked in such cold. I'm all for saving the planet, but this was barbaric." Mr. Benzie declared the day a success and plans similar "eco-days" in the future.[38]

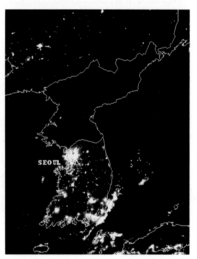

It's Earth Hour every night for the 22 million people of North Korea. (NASA)[40]

On a global scale, the prize for the most futile climate gesture goes to Earth Hour. Earth Hour was first organized by the World Wildlife Fund in 2007 and is now an annual event on March 31. Climatists urge households and businesses all over the world to turn off their lights for one hour to "dare the world to save the planet."[39]

What must the roughly 1.5 billion people of the world who don't have electricity think about Earth Hour? People without lights, appliances, computers, proper health care, and modern industry, all made possible by electricity, probably question the wisdom of turning off the lights. Maybe the Climatists have their priorities a bit fouled up.

WACKY CLIMATE CHANGE HEADLINES

Climate Change Could Be Causing Cougar Attacks: Expert
—*Canada.com*, Aug. 29, 2007[41]

ANIMAL SCIENTISTS LOVE THIS WARMING STUFF

Animal scientists love global warming. Governments are happy to fund scientific studies on man-made impacts to the biosphere. Biologists, ecologists, health experts, ichthyologists, ornithologists, veterinarians, and researchers from all over the world receive funding to report on how man-made global warming is impacting the plant and animal kingdom.

Climate Change Cited as Shark Attacks 'Double'
—*The Raw Story*, February 13, 2012[42]

They tell us that elephants, frogs, orangutans, polar bears, tigers, and the duck-billed platypus are all threatened by global warming.[43,44] Many of the studies claim that species that have been around for millions of years (and therefore have survived the 7°C–12°C temperature swings of the ice ages) are now threatened by the 1°C warming over the last 150 years.

Some climate change headlines are downright bizarre. We're warned that vampire bats are killing more children in Brazil,[45] that Lyme disease-carrying ticks in Canada are flourishing,[46] that a new species of mutant sharks has been discovered in Australia,[47] and

Lizard that Outlived Dinosaurs May Go Extinct from Climate Change
—*Planet Save*, June 19, 2010[48]

that tropical fish of the Great Barrier Reef are losing their hearing[49]—all attributed to man-made warming. Scientists in Malaysia even found that local flowers were losing their scent. On expert advice, the Kuala Lumpur City Hall changed its planting focus from flowers to trees "more suited for soaking up the increasing pollution and coping with global warming."[50]

CLIMATE CHANGE FOOLISHNESS

Yes, the very learned and clever can be foolish when it comes to climate change. From green, leafy fighter jets to camel hunts from helicopters to zero-carbon classrooms that freeze students, humanity provides abundant examples of zany behavior in a futile effort to fight global warming. But it appears that Climatism is headed for a fall.

CHAPTER 13

CLIMATISM—HEADED FOR A CRASH

"It ain't what you don't know that gets you into trouble.
It's what you know for sure that just ain't so."
—AUTHOR MARK TWAIN

For more than 20 years, the Intergovernmental Panel on Climate Change (IPCC) and the supporting scientific community, led by climate computer modelers, have successfully promoted the theory of man-made global warming. Almost every nation has accepted prophesies of disaster, responding with bizarre programs of every kind. Researchers at universities and government laboratories have cried "Alarm! Alarm!," gathering billions in funding to "solve" the climate crisis. Vast sums have been spent on dilute, intermittent, and

237

expensive renewable energy projects—money that could be used instead to solve the real pressing problems of mankind. Businesses both large and small promote sustainability and sell green products, trusting that they are helping the environment. But as we discussed in Chapters 4 and 5, climate change is due to natural processes, probably driven by the sun, and man-made emissions play only a very small part. So the world is living in climate madness, certain of a pending climate catastrophe that isn't going to happen.

Climatism is headed for a fall. Recent trends in scientific data indicate the IPCC projections of catastrophe appear more and more far-fetched. Political leaders, once marching down the road of Climatism, are no longer certain of the right path. The renewable energy revolution is now threatened by the hard realities of economics and a new abundance of hydrocarbon energy. And citizens are questioning the scary scenario of man-made global warming. The coming disaster will not be about Earth's climate, but will instead be the destruction of the theory of man-made global warming.

TRENDS SHOW THAT THE MODELS ARE WRONG

We've now had 20 years to assess IPCC projections, and the projections are found wanting. Based on climate model simulations, the IPCC 1990 First Assessment Report told the world to expect a temperature rise of 0.3°C per decade, leading to 2025 temperatures that would be 1°C higher than 1990 temperatures.[1] But satellite data shows that global temperatures have been flat to declining over at least the last ten years. Temperatures remain about 0.2°C above 1990 levels, but the rise has been substantially lower than the IPCC "low estimate" projection.[2]

In a 2005 paper, Dr. James Hansen, head of NASA's Goddard Institute, warned of an "energy imbalance" due to rising atmospheric CO_2 that was causing Earth's oceans to absorb energy. He stated that this was confirmed by "precise measurements over the last 10 years."[3] But measurements of ocean heat content before 2005 were based on sporadic data taken by passing ships, which was anything but precise.

Since 2000, 24 nations and the European Union have cooperated to deploy the Argo network. Argo is a global array of 3,500 buoys that measure temperature and salinity of the upper 2,000 meters of the ocean.[4] With better ocean coverage and consistent measurements, Argo shows a surprising result. For the last eight years, there has been *no change* in ocean heat content, despite climate model predictions of a steady heat content rise. Dr. Kevin Trenberth of the National Center for Atmospheric Research was baffled by

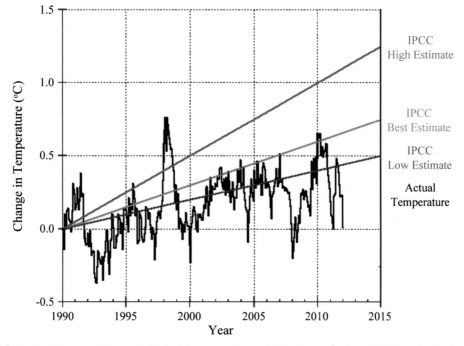

IPCC Projections and Actual Global Temperature. High, best, and low IPCC projections for global temperature from the IPCC First Assessment Report in 1990, compared to actual global temperature from satellite data. (IPCC First Assessment Report, 1990; UAH Satellite Data; Evans, 2012)[5,6]

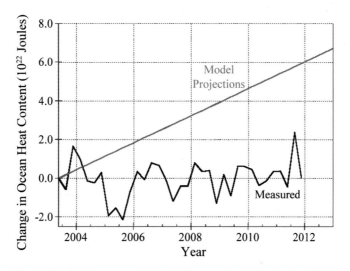

Model Projections and Actual Ocean Heat Content. Model projections diverging from Argo buoy measurements. (Hansen, 2005; National Oceanic Data Center; Evans, 2012)[7,8]

the "missing heat," stating in a 2009 paper:

> …Was it because a lot of the heat went into melting Arctic sea ice or parts of Greenland
> and Antarctica, and other glaciers? Was it because the heat was buried in the ocean and
> sequestered, perhaps well below the surface?…Perhaps all of these things are going on?[9]

In addition to the actual measurements of surface temperatures and ocean heat, natural climate cycles appear to be moving against the theory of man-made warming. Within the last ten years, two of the Earth's major climate cycles, the Pacific Decadal Oscillation and the Atlantic Multidecadal Oscillation, appear to have moved into a cool phase. Many solar physicists now predict a period of low solar activity in the current and coming Sunspot Cycles 24 and 25.[10] All of these trends point to a coming period of cooler global temperatures, rather than the feared warming.

SCIENCE DOES NOT SUPPORT CLIMATE ALARM

After 20 years, it's clear that climate catastrophe is not occurring. The modest 0.5°C warming from 1975 to 2000 was not abnormal compared to the Medieval Warm Period and other eras, and there has been no global warming for the last ten years. There is no evidence of increasing water vapor in Earth's atmosphere, and the model-predicted hot spot over the tropics has not appeared.

Water vapor is Earth's most abundant greenhouse gas. Contributions of carbon dioxide to Earth's atmosphere are overwhelmingly due to releases from the oceans, biosphere, and volcanoes. Man-made emissions produce only about one percent of Earth's greenhouse effect. Carbon dioxide is plant food and the best substance humanity could release into the biosphere.

The Antarctic Icecap, which contains 90 percent of Earth's ice, continues to slowly expand. Ocean levels are rising at seven to eight inches per century, not the 20 feet per century predicted by Dr. Hansen and Mr. Gore. Hurricanes and tornados are neither more frequent nor more powerful on a global scale than those of the past. Polar bear populations are at a 50-year high, and stories of the bear's demise are greatly exaggerated. The fear of ocean acidification is based solely on computer model projections, without empirical evidence that changes in ocean pH are historically abnormal.

Climate science jumped to a wrong conclusion more than 20 years ago, and Climatism is now driven by money. The Climategate emails revealed that lead authors of the IPCC reports were strongly biased to develop data to support the false theory of man-made global warming. Despite the mounting evidence, NOAA, NASA, National Academy of Sciences, the Royal Society, and all major scientific organizations of the world continue to support the theory of man-made global warming. Many will be dining on a generous helping of crow when the world returns to climate reality.

THE FAILURE OF GLOBAL NEGOTIATIONS

The year 2009 was set to be a year of triumph for Climatism. In 2007, the IPCC's Fourth Assessment Report declared that mankind was very likely the cause of global temperature increase.[11] That same year, Al Gore and the IPCC shared the Nobel Peace Prize. In 2008, Barack Obama was elected President of the United States, heralding a rebirth of a more environmentally conscious nation. After securing the majority of primary delegates in June 2008, candidate Obama declared:

> …this was the moment when the rise of the oceans began to slow and our planet began to heal…[12]

Following the President's lead, the US House of Representatives passed the Waxman-Markey cap-and-trade bill in June 2009 and sent it to the US Senate.[13]

The Copenhagen Conference in December 2009, part of the United Nations Framework Convention on Climate Change (UNFCCC), was to be the major next step to control global emissions. Climate activists called for a successor treaty to replace and expand the Kyoto Protocol and establish binding emissions limits on all countries. In January 2009, the European Community (EC) proposed a 30-percent emissions cut by year 2020 for developed nations and a 15- to 30-percent reduction in emissions from "business as usual" levels by large developing nations.[14] More than 15,000 conference attendees, including President Obama and more than 100 other heads of state, traveled to Copenhagen. Despite differences between the developed and developing nations, conference delegates were cautiously optimistic.

But then the momentum collapsed. Throughout 2008 and 2009, opposition to the theory of man-made global warming gained headway with

"We are disillusioned. The current political system is broken…Essentially nothing has changed in 20 years. We are not remotely on a course to be sustainable."
—Robert Watson, UK Chief Environmental Advisor, former IPCC Chairman (2012)[15]

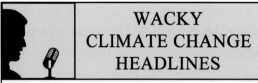

WACKY CLIMATE CHANGE HEADLINES

China Threatens Massive Venting of Super Greenhouse Gases in Attempt to Extort Billions as UNFCCC Meeting Approaches

"China has responded to efforts to ban the trading of widely discredited HFC-23 offsets by threatening to release huge amounts of the potent industrial chemical into the atmosphere unless other nations pay what amounts to a climate ransom."
—*PR Newswire*, Nov. 8, 2011[16]

world opinion. The Heartland Institute sponsored the first of eight International Conferences on Climate Change in 2008 and 2009, providing skeptical scientists with a forum for realist climate views.[17] In 2009, the Nongovernmental International Panel on Climate Change issued *Climate Change Reconsidered*, an 880-page volume that was the most extensive critique of the IPCC at the time. The report cited thousands of peer-reviewed articles and concluded that "natural causes are very likely to be the dominant cause" of global warming.[18] Citizens began to recognize that man-made warming alarm was a political movement.

Then in late November 2009, just one week prior to the Copenhagen summit, the release of the Climategate emails shook the science of man-made warming. On their home turf of Copenhagen, European delegates had intended to use the conference to lead the world to a more aggressive climate treaty. But they were shocked when their proposals were ignored by developing nations. On the last day, Brazil, China, India, South Africa, and the United States crafted a weak, voluntary agreement named the Copenhagen Accord, which was then adopted.[19] Delegates left the conference without either a binding agreement on emissions limits or a successor treaty to the Kyoto Protocol.

The combination of the failure at Copenhagen and the release of the Climategate emails shook many in the world community. German Chancellor Angela Merkel and other European leaders returned from the conference disillusioned and discouraged about

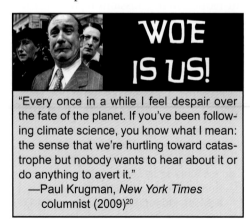

WOE IS US!

"Every once in a while I feel despair over the fate of the planet. If you've been following climate science, you know what I mean: the sense that we're hurtling toward catastrophe but nobody wants to hear about it or do anything to avert it."
—Paul Krugman, *New York Times* columnist (2009)[20]

the prospects for a global treaty.[21] Copenhagen shattered the illusion that the world would join a Europe-led climate crusade.

In 2010, more bad news arrived from the United States. First, the Waxman-Markey cap-and-trade bill died in the Senate without even being called to a vote. Then in fall of 2010, mid-term elections saw Republicans capture a majority in the House of Representatives and gain seats in the Senate. Many first-term

Republican representatives openly challenged the theory of man-made warming.

The 2010 Cancun and 2011 Durban climate conferences did little to restore the stalled momentum of the global warming movement. At Cancun, Mexico, representatives reaffirmed their commitment to limit the global temperature rise to 2°C and their pledge to establish a $100-billion climate fund for developing nations.[22] Not that there was any new evidence that mankind could control global temperatures. The Durban, South Africa, conference called on members to negotiate a new agreement on binding emissions that would include all nations, to go into effect by 2020. The conference also proposed an extension of the Kyoto Protocol for another five years.[23]

But key delegates to Durban were not pleased. Representatives from China and India made it clear that they were unhappy with emissions restrictions, stating that the industrial nations were responsible for emissions that caused the global warming problem. Shortly after the end of the Durban conference, Canada announced that it would not participate in an extension of the Kyoto Protocol. Japan and Russia subsequently also declined to participate, and the United States repeated that it would remain outside of the treaty.[24]

After 20 years of climate negotiations and millions of hours of delegate time, the misguided nations of the world have achieved little. The Kyoto treaty ends in 2012, and a wide policy position gap continues to exist between the developed and developing nations. By 2010, global greenhouse gas emissions were *up 45 percent* from 1990 levels. Despite massive efforts to convert to renewables and to enact cap-and-trade and other foolish climate laws, *the growth rate in global emissions from 1990 to 2010 was the same as the growth rate from 1970 to 1990.*[25] It's no wonder most Climatists are in despair.

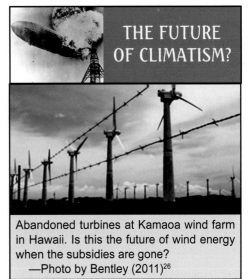

THE FUTURE OF CLIMATISM?

Abandoned turbines at Kamaoa wind farm in Hawaii. Is this the future of wind energy when the subsidies are gone?
—Photo by Bentley (2011)[26]

RENEWABLE REMEDIES ARE BANKRUPT

It's increasingly clear that the remedies of Climatism have failed. As we discussed in Chapter 11, wind and solar cannot replace conventional power plants if continuity of electrical

THE FUTURE
OF CLIMATISM?

After only two years, this solar farm in Markranstadt in eastern Germany is overgrown with weeds.
—Photo by Loschke (2011)[27]

supply is to be maintained. As we discussed in Chapter 11, the use of wind turbines forces backup plants to cycle inefficiently, releasing more harmful pollutants and CO_2 emissions than conventional power plants without wind. Similarly, when land usage is taken into account, vehicles using biofuel produce more pollutants and CO_2 emissions than gasoline- or diesel-powered vehicles. "Sustainable" renewable sources use dozens of times more land than hydrocarbon or nuclear alternatives, and the production of biofuels requires much more water than gasoline or diesel fuel.

Rather than being at a point of peak oil and gas, we may be at a point of peak renewables, at least as a percentage of global energy usage. The great hope of Climatism was that the cost of hydrocarbons would continue to rise, making renewables competitive. But the hydraulic fracturing revolution has driven the cost of natural gas down 70 percent in the US over the last decade. Electricity from natural gas plants is cheaper, generates lower emissions than combined wind-gas systems, and has a much smaller land footprint than wind systems. It appears that fracking will provide mankind with a supply of low-cost gas for at least 200 years. Why subsidize another wind turbine?

Faced with massive commitments for renewable subsidies, Germany, Greece, the Netherlands, Spain, the United Kingdom, and the United States have all recently cut subsidies for renewables. How many subsidies will remain when people realize that humans are not destroying Earth's climate? When the subsidies disappear, will fields and hills remain scarred by rusting turbines, with wind howling past unturning blades? Who will clean up the acres of solar cell panels, broken and weed-infested?

EUROPE AND DECARBONIZATION FOLLY

Europe is living in a fantasy world. The EC continues to push for a zero-carbon future, calling for an 80-percent emissions reduction by the year 2050. Although 2010 CO_2 emissions from 27 European nations were down 7 percent since 1990,[28] much of this is due to a shift from production to imports. For example, official UK government numbers showed a 22-percent decline in CO_2 emissions from UK industry from 1990 to 2009. But total CO_2 emissions associated with UK consumption *rose* 12 percent over the same period,[29] because more goods were imported from abroad. When imports are considered, European nations aren't cutting anything.

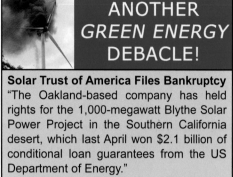

ANOTHER GREEN ENERGY DEBACLE!

Solar Trust of America Files Bankruptcy
"The Oakland-based company has held rights for the 1,000-megawatt Blythe Solar Power Project in the Southern California desert, which last April won $2.1 billion of conditional loan guarantees from the US Department of Energy."
—*Thomson Reuters*, April 2, 2012[30]

Energy economics will preclude deep emissions cuts short of economic destruction. Solar systems don't deliver enough energy to make a difference. Wind systems, when considered with required hydrocarbon backup, don't reduce emissions. Nuclear, the only viable decarbonization choice, has been rejected by Austria, Belgium, Denmark, Germany, Italy, and Spain.

Transportation poses a special problem for decarbonization. European emissions from the transportation sector increased 36 percent from 1990 to 2007.[31] As we discussed in Chapter 11, studies now

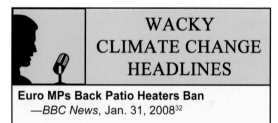

WACKY CLIMATE CHANGE HEADLINES

Euro MPs Back Patio Heaters Ban
—*BBC News*, Jan. 31, 2008[32]

show that ethanol and biodiesel fuels *do not* reduce emissions when used in place of gasoline, diesel, and aviation fuels. The EC has no alternative but to stop the use of cars, trucks, and planes, if transportation emissions are to be cut.

Nevertheless, Europe continues to march down the road of climate madness. Even a patio heater ban has been proposed to halt climate change. When will European citizens wake up to reality?

SUSTAINABILITY: A HOUSE BUILT ON SAND

According to the United Nations, "sustainable development" is development that meets "the needs of the present without compromising the needs of the future."[33] Sustainability is now widely accepted by governments, businesses, universities, and most major organizations.

Economics taught us that resource scarcity was resolved by market pricing through supply and demand and substitution of goods. But sustainability tells us this isn't good enough. Instead, intellectual elites must direct the activities of mankind to preserve the planet for future generations.

At the core of sustainable development is a belief in man-made global warming. The United Nations and other proponents of sustainability call for reduced consumption, reduced production, and reduced energy usage. Wind, solar, and biofuels are defined as "sustainable," despite their ineffectiveness and land-hogging qualities when compared to traditional fuels. But since nature, not man, controls the climate, the philosophy of sustainable development is built on falsehood.

ONE TRILLION DOLLARS DOWN A GREEN DRAIN

Every day, 25,000 people die from hunger-related issues in developing nations. More than 1 billion people are trying to survive on less than $1.25 per day. Two and one-half billion people do not have adequate sanitation, 1.4 billion do not have electricity, and almost 1 billion do not have access to clean drinking water. Every year, 2 million die from AIDS. Almost 2 million die from tuberculosis. Malaria, pneumonia, and diarrheal diseases kill millions more.[34]

The tragedy of Climatism is a misuse of resources on a vast scale. The world spent $243 billion in 2010 on renewable energy, trying to "decarbonize" energy systems.[35] *More than $1 trillion was spent over the last ten years*, and governments and industries are on pace to waste another $1 trillion in the next four years on foolish climate programs. Each year, twice as much is spent in a futile attempt to stop global warming as is spent for total international aid.[36] Imagine the benefits to the world's poor if decarbonization expenditures could be redirected to solve the problems of hunger, disease, and poverty.

BILLIONS WILL FIGURE IT OUT

Today, billions of people believe in the theory of man-made global warming. But year after year, temperatures do not follow model predictions, sea levels do not rise abnormally, the polar bears thrive, and predicted disasters do not occur. The world's citizens will figure it out. Changes in public opinion already show that citizens are beginning to learn the real story. The crash of Climatism will be thunderous.

Let's hasten the fall of Climatism and the awakening of mankind to climate reality. Climate change is natural and cars are innocent. Let's reallocate the vast funds spent in foolish efforts to fight global warming, to instead solve the real pressing problems of mankind.

FURTHER READING

Blue Planet in Green Shackles: What is Endangered: Climate or Freedom? by Václav Klaus (Competitive Enterprise Institute, 2008)

Climate Change Reconsidered: the Report of the Nongovernmental International Panel on Climate Change by Craig Idso and S. Fred Singer (Heartland Institute, 2009)

Climategate: A Veteran Meteorologist Exposes the Global Warming Scam by Brian Sussman (WND Books, 2010)

Climate Confusion: How Global Warming Hysteria Leads to Bad Science, Pandering Politicians and Misguided Policies that Hurt the Poor by Roy Spencer (Encounter, 2010)

Climate Coup: Global Warming's Invasion of Our Government and Our Lives by Patrick J. Michaels (Cato Institute, 2011)

Climate of Corruption: Politics and Power Behind the Global Warming Hoax by Larry Bell (Greenleaf Book Group, 2011)

Climate: The Counter Consensus by Robert M. Carter (Stacey International, 2010)

Climatism! Science, Common Sense, and the 21st Century's Hottest Topic by Steve Goreham (New Lenox Books, 2010)

CO_2, Global Warming and Coral Reefs: Prospects for the Future by Craig Idso (Vales Lake Publishing, 2009)

Eco-Imperialism: Green Power, Black Death by Paul Driessen (Merril Press, 2010)

Heaven and Earth: Global Warming, the Missing Science by Ian Plimer (Taylor Trade, 2009)

Power Hungry: The Myths of "Green Energy" and the Real Fuels of the Future by Robert Bryce (Public Affairs, 2011)

Power Politics by Michael J. Economides and Peter C. Glover (ET Publishing, 2012)

The Chilling Stars: A New Theory of Climate Change by Henrik Svensmark and Nigel Calder (Totem, 2003)

The Delinquent Teenager Who Was Mistaken for the World's Top Climate Expert by Donna Laframboise (Createspace, 2011)

The Deniers: The World Renowned Scientists Who Stood Up Against Global Warming Hysteria, Political Persecution, and Fraud—And those who are too fearful to do so by Lawrence Solomon (Richard Vigilante Books, 2008)

The Greatest Hoax: How the Global Warming Conspiracy Threatens Your Future by Senator James Inhofe (WND Books, 2012)

The Great Global Warming Blunder: How Mother Nature Fooled the World's Top Climate Scientists by Roy W. Spencer (Encounter Books, 2010)

The Politically Incorrect Guide to Global Warming (and Environmentalism) by Christopher C. Horner (Regnery, 2007)

The Real Global Warming Disaster by Christopher Booker (Continuum, 2010)

The Skeptical Environmentalist: Measuring the Real State of the World by Bjorn Lomborg (Cambridge University, 2001)

The Ultimate Resource 2 by Julian L. Simon (Princeton University Press, 1998)

Unstoppable Global Warming: Every 1500 Years by S. Fred Singer and Dennis Avery (Rowman & Littlefield, 2007)

Watermelons: The Green Movement's True Colors by James Delingpole (Publius Books, 2011)

NOTES

Introduction

1. Peter Gwynne, "The Cooling World," *Newsweek*, April 28, 1975, p.64; Crystal ball image by Eva Kröcher under GFDL
2. Photograph by Dawn Guenther, all rights reserved
3. "EPA Awards $17 Million to Support Research on the Impacts of Climate Change Twenty-five universities to explore public health and environmental facets of climate change," EPA press release, Feb. 2, 2010, http://yosemite.epa.gov/opa/admpress.nsf/e77fdd4f5afd88a3852576b3005a604f/806e135c0522699b852576cd006b4813!OpenDocument

Chapter 1: Mankind in the Grip of a Madness

1. "The Nobel Peace Prize for 2007," *Nobelprize.org*, http://nobelprize.org/nobel_prizes/peace/laureates/2007/press.html
2. Al Gore photograph by Kjetil Bjørnsrud under GNU Free Documentation License
3. Al Gore, The Nobel Lecture, Oslo, Dec. 10, 2007, http://nobelpeaceprize.org/enGB/laureates/laureates-2007/gore-lecture/
4. Intergovernmental Panel on Climate Change, http://www.ipcc.ch/organization/organization.shtml
5. The Nobel Peace Prize 2007, http://nobelprize.org/nobel_prizes/peace/laureates/2007/ipcc-lecture_en.html
6. HULIQ, http://www.huliq.com/44398/un-climate-panel-chief-pachauri-gore-accept-nobel-peace-prize
7. *An Inconvenient Truth: The Planetary Emergency of Global Warming and What We Can Do About It* by Al Gore (Rodale, 2006)
8. *Our Choice: A Plan to Solve the Climate Crisis by Al Gore* (Rodale, 2009)
9. Al Gore, statement to the Senate Foreign Relations Committee, January 28, 2009
10. Image spoof by Anthony Watts
11. Al Gore image by US Government Images, http://globalwarming.house.gov /tools/assets/files/0127.jpg; James Hansen image by Bill Ebbesen; Bill McKibben image by Evan Derickson under GFDL; Nicholas Stern image by the International Monetary Fund; Tim Flannery image by Mark Coulson, 5th World Conference of Science Journalists
12. Al Gore (Rodale, 2006), p. 10.
13. James Hansen, letter to Barack and Michelle Obama, December 29, 2008
14. *Eaarth: Making a Life on a Tough New Planet* by Bill McKibben (Times Books, 2010), p. 2
15. Alison Benjamin, "Stern: Climate Change a 'Market Failure,'" *Guardian*, November 29, 2007
16. *Now or Never: Why We Must Act Now to End Climate Change and Create a Sustainable Future* by Tim Flannery (Atlantic Monthly Press, 2009), p. 14
17. James Hansen, "Coal-fired power stations are death factories. Close them," *Observer*, February 15, 2009, http://www.guardian.co.uk/commentisfree/2009/feb/15/james-hansen-power-plants-coal
18. J. Hansen et. al., "Climate Impact of Increasing Atmospheric Carbon Dioxide," *Science*, Aug. 28, 1981, Vol. 213, No. 4511
19. James Hansen, "Global Warming Twenty Years Later: Tipping Points Near," address to the National Press Club, June 232, 2008; Executioner photo by Andrew Butko under GFDL
20. *Eaarth* (See no. 14)
21. "*350.org* coordinated 5,200 rallies" *350.org* website: http://www.350.org/en/mission
22. Andrew Revkin, "The Long-Distance Climate Campaigner," *The New York Times*, December 5, 2010
23. *The Economics of Climate Change: The Stern Review* by Nicholas Stern (Cambridge University Press, 2007)
24. "Mass Migrations and War: Dire Climate Scenario," *Homeland Security News*, February 24, 2009, http://homelandsecuritynewswire.com/dire-climate-scenario-mass-migrations-and-war

25. Ibid

26. Barbara Hollingsworth, "Spend trillions now, and world temperatures might fall in 1,000 years," *Washington Examiner*, Mar. 26, 2011, http://washingtonexaminer.com/blogs/beltway-confidential/2011/03/spend-trillions-now-and-world-temperatures-might-fall-1000-years

27. *The Weather Makers: How Man is Changing the Climate and What It Means for Life on Earth* by Tim Flannery (Grove Press, 2001) p.17

28. "Bill Clinton, Al Gore Get Rich After White House," *ABC News*, June 15, 2007, http://abcnews.go.com/GMA/story?id=3281925

29. David Cameron image by 10 Downing Street under Open Government License; Barack Obama image by Pete Souza under Creative Commons Unported License; Julia Gillard image by MystifyMe Concert Photography (Troy) licensed under Creative Commons Generic License; Angela Merkel image by Agência Brasil under Creative Commons License Brazil; Manmohan Singh image by José Cruz under Creative Commons Brazil license

30. David Cameron, April 8, 2010, http://www.youtube.com/watch?v=HjU0CQKQxPo

31. Barack Obama, Message for the Global Climate Summit, November 17, 2008

32. "Australian prime minister lays out climate change plan" *CNN*, July 23, 2010, http://articles.cnn.com/2010-07-22/world/australia.climate.change_1_climate-change-carbon-pollution-biggest-polluters?_s=PM:WORLD

33. Angela Merkel, UN Climate Change Summit, September 22, 2009

34. Mahoman Singh, speech at Council on Foreign Relations, November 23, 2009

35. Debbie Stabenow, interview with *Detroit News*, Aug. 8, 2009, http://community.detnews.com/apps/blogs/henrypayneblog/index.php?blogid=2041; Hot earth image by Jack1 under GFDL

36. "Who's Winning the Clean Energy Race," The Pew Charitable Trusts, 2011, http://www.pewtrusts.org/uploadedFiles/wwwpewtrustsorg/Reports/Global_warming/G-20%20Report.pdf

37. US Conference of Mayors Climate Protection Agreement, http://www.usmayors.org/climateprotection/agreement.htm

38. C40 Cities Climate Leadership Group, http://live.c40cities.org/about-us/

39. American College & University Presidents' Climate Commitment, http://presidentsclimatecommitment.org/about/commitment

40. Christian Hiller, "Deutsche Bank, Al Gore and the $10 Billion Climate Fund," *Frankfurter Allgemeine Zeitung*, November 15, 2010, http://www.thegwpf.org/international-news/1860-deutsche-bank-al-gore-and-the-10-billion-climate-fund.html

41. "European Consumers Pay 46% More for Green Retail Products," *Kelkoo*, May 31, 2010, http://press.kelkoo.co.uk/european-consumers-pay-46-more-for-green-retail-products.html.

42. Natural Marketing Institute, "LOHAS Market Size," *LOHAS Journal*, Spring 2010, http://www.lohas.com/sites/default/files/lohasmarketsize.pdf

43. Ashley Cleek, "Russian Scholar Warns Of 'Secret' US Climate Change Weapon," *RadioFreeEuropeRadioLiberty*, July 30, 2010, http://www.rferl.org/content/Russian_Scholar_Warns_Of_Secret_US_Climate_Change_Weapon/2114381.html

44. "World's Largest Investors, Worth $20 Trillion, Step Up Call for Urgent Policy Action on Climate Change, The Corporate Social Responsibility Newswire, Oct. 20, 2011, http://www.csrwire.com/press_releases/33186-World-s-Largest-Investors-Worth-20-Trillion-Step-Up-Call-for-Urgent-Policy-Action-on-Climate-Change

45. US Climate Action Partnership, http://www.us-cap.org/

46. Peter T. Doran and Maggie Kendall Zimmerman, "Examining the Scientific Consensus on Climate Change," *EOS, Trans. AGU*, 90(3), October 29, 2009

47. American Physical Society, http://www.aps.org/policy/statements/07_1.cfm

48. National Atmospheric and Oceanographic Administration, http://www.ncdc.noaa.gov/indicators

49. National Aeronautic and Space Administration, http://climate.nasa.gov/evidence

50. American Geophysical Union, http://www.agu.org/sci-pol/positions/climate_change2008.sh

51. American Meteorological Society, http://www.ametsoc.org/policy/2007climatechange.html

52. National Academy of Sciences, http://dels.nas.edu/report/america-climate-choices-2011/12

53. "Climate Change: A Summary of the Science," The Royal Society, Sep. 2010

54. Royal Meteorological Society, http://www.rmets.org/about/press_detail.php?ID=33

55. "Let's Be Honest," European Academy of Arts and Sciences, http://www.euro-acad.eu/downloads/memorandas/lets_be_honest_-_festplenum_03.03.07_-_final2.pdf

56. Letter to the Canadian Parliament, Canadian Meteorological and Oceanographic Society et al., November 26, 2009

57. Deborah Amos, "Hot Politics," *PBS Frontline*, Apr. 24, 2007, http://www.pbs.org/wgbh/pages/frontline/hotpolitics/etc/script.html

58. NASA photograph, http://en.wikipedia.org/wiki/File:James_Hansen.jpg

59. UN Framework Convention on Climate Change, May 9, 1992, http://unfccc.int/resource/docs/convkp/conveng.pdf

60. Statement by Atmospheric Scientists on Greenhouse Warming," February 27, 1992, http://www.his.com/~sepp/policy%20declarations/statment.html

61. United Nations Framework Convention on Climate Change, http://unfccc.int/kyoto_protocol/items/2830.php

Chapter 2: Do This to Save the Planet

1. Audra Ang, "Pelosi appeals for China's help on climate change," *Associated Press*, May 27, 2009, http://www. thefreelibrary.com/Pelosi+appeals+for+China's+help+on+climate+change-a01611879983
2. UN Framework Convention on Climate Change, http://unfccc.int/files/meetings/cop_16/application/pdf/cop16_lca.pdf
3. White House Press Release, November 25, 2009, http://www.whitehouse.gov/the-press-office/president-attend-copenhagen-climate-talks
4. *Ecoscience: Population, Resources, Environment* by Paul R. Ehrlich, et. al. (W. H. Freeman, 1970), p. 323
5. N.C. Aizenman, "U.S. to Grow Grayer, More Diverse," *Washington Post*, Aug. 14, 2008, http://www.washingtonpost.com/wp-dyn/content/article/2008/08/13/AR2008081303524_pf.html
6. Statistisches Bundesamt Deutschland, Nigeria, http://www.destatis.de/jetspeed/portal/cms/Sites/destatis/Internet/EN/Content/Statistics/Internationales/InternationalStatistics/Country/Africa/Nigeria,templateId=renderPrint.psml
7. Cell phone image by Kristoferb under Creative Commons License; Personal computer image by Boffyb, under GFDL; Jet aircraft image by Adrian Pingstone; Television image by Varglko; Automobile image by Luna04 under GFDL; Light bulb image by KMJ under GFDL
8. Emissions graph data from Carbon Dioxide Information Analysis Center, http://cdiac.ornl.gov/trends/emis/tre_usa.html
9. US Census Bureau, http://www.census.gov/prod/www/abs/statab.html
10. *Climatism! Science Common Sense, and the 21st Century's Hottest Topic* by Steve Goreham (New Lenox Books, 2010), p. 268-275
11. Ibid, p. 268-275
12. IEA Wind 2010 Annual Report, p. 5-6, http://www.ieawind.org/
13. Key World Energy Statistics, International Energy Agency, 2011, http://www.iea.org/textbase/nppdf/free/2011/key_world_energy_stats.pdf
14. *Climatism!* (See no. 10), p. 286
15. *Climatism!* (See no. 10), p. 278
16. *Climatism!* (See no. 10), p. 285
17. Ronald Steenblik, "Biofuels—At What Cost?," Global Subsidies Initiative of the International Institute for Sustainable Development, September, 2007, http://www.globalsubsidies.org/en/research/biofuel-subsidies-selected-oecd-countries
18. Al Gore, Repower America, 2009, http://www.youtube.com/watch?v=xrUckZXQcAQ "Renewable Fuel Standard, Potential Economic and Environmental Effects of US Biofuel Policy," National Research Council, 2011, p. 221
19. Key World Energy Statistics (See no. 13)
20. IEA Wind (See no. 12)
21. *The First Global Revolution: A Report by the Council of The Club of Rome* by Alexander King and Bertrand Schneider (Pantheon, 1991), p. 115; Spy image by Setreset under Creative Commons Attribution License
22. "Green Transportation Solutions: Clean Cars," Sierra Club, http://www.sierraclub.org/transportation/cleancars/default.aspx
23. "Passenger Vehicle Greenhouse Gas and Fuel Economy Standards: A Global Update," International Council on Clean Transportation, July, 2007, http://www.lowcvp.org.uk/assets/reports/ICCT_GlobalStandards_2007.pdf
24. Ibid
25. "What are the costs of CO_2 Reductions through Vehicle Technology?" European Automobile Manufacturers Association, January 2007, http://www.acea.be/images/uploads/co2/Cost-analyses_European_ Climate_Change_Programme.pdf.
26. John M. Broder, "Obama to Toughen Rules on Emissions and Mileage," *New York Times*, May 19, 2009, http://www.nytimes.com/2009/05/19/business/19emissions.html
27. Letter to President Obama, September 9, 2010, http://www.ucsusa.org/assets/documents/clean_vehicles/Letter-to-Obama-to-Increase-Fuel-Efficiency-and-Cut-Tailpipe-Pollution.pdf.
28. "2017 and Later Model Year Light-Duty Vehicle Greenhouse Gas Emissions and Corporate Average Fuel Economy Standards," Environmental Protection Agency and Department of Transportation, *Federal Register*, v. 76, no. 231, Dec. 1, 2011, http://www.nhtsa.gov/About+NHTSA/Press+Releases/2011/We+Can't+Wait:+Obama+Administration+Proposes+Historic+Fuel+Economy+Standards+to+Reduce+Dependence+on+Oil,+Save+Consumers+Money+at+the+Pump
29. Horse and carriage image by Steve Evans under Creative Commons Attribution License; Leaf image by uneekGrafix under GNU General Public License
30. ICCT (See no. 23)
31. "Best Practices for Feebate Design and Implementation," International Council on Clean Transportation, May, 2010, http://www.theicct.org/pubs/feebate_may10.pdf
32. Margaret Zewatsky, "What Drives Global Hybrid Sales," March 31, 2010, http://blog.polk.com/blog/blog-posts-by-margaret-zewatsky/0/0/what-drives-global-hybrid-sales
33. "Study: A Dog or Cat Pollutes for than an SUV," *Italian News Agency*, August 18, 2010, http://hockeyschtick.blogspot.com/2010/09/study-dog-or-cat-pollutes-more-than-suv.html
34. "Barack Obama and Joe Biden: New Energy for America," August 3, 2008, http://www.barackobama.com/pdf/factsheet_energy_speech_080308.pdf
35. Sharon Silke Carty, "Obama Pushes Electric Cars, Battery Power This Week," *USA Today*, July 14, 2010, http://content.usatoday.com/communities/driveon/post/2010/07/obama-pushes-electric-cars-battery-power-this-week-/1

36. "UK to Produce Nissan Electric Car," *BBC News*, March 18, 2010, http://news.bbc.co.uk/2/hi/business/8573724.stm

37. John Reed, "Buyers Loath to Pay more for Electric Cars," *Financial Times*, Sep. 19, 2010, http://www.ft.com/cms/s/0/acc0a646-c405-11df-b827-00144feab49a, dwp_uuid=aece9792-aa13-11da-96ea-0000779e2340.html#axzz1DxCyYD8Q

38. "David Cameron: The Green Consumer Revolution, The Conservative Party, October 16, 2009, http://www.conservatives.com/News/Speeches/2009/10/David_Cameron_The_ Green_Consumer_Revolution.aspx

39. House of Commons Hansard Ministerial Statements, July 27, 2010, http://www.publications.parliament.uk/pa/cm201011/cmhansrd/cm100727/wmstext/100727m0001.htm.

40. "Carbon Compliance: What is the Appropriate Level for 2016," Zero Carbon Hub, November, 2009, http://www.zerocarbonhub.orgresourcefiles /Carbon_Compliance_Interim_Report_16_12_10.pdf

41. "What to do if a Compact Fluorescent Light (CFL) Bulb or Fluorescent Tube Light Bulb Breaks in Your Home," US Environmental Protection Agency, Dec. 1, 2010, http://www.deq.state.va.us/export/sites/default/recycle/CFLcleanupguidanceEPA2010.pdf

42. Victoria Allen and Tamara Cohen, "Climate Minister buys a castle with 16 bathrooms…and a massive carbon footprint," *Daily Mail*, Nov. 18, 2011, http://www.dailymail.co.uk/news/article-2063488/Climate-Minister-buys-castle-16-bathrooms--massive-carbon-footprint.html

43. Light bulb image by KMJ under GNU Free Documentation License

44. Michael O'Brien, "Coalition to Tackle Labor's Smart Meter Stuff Up, Liberal Victoria," November 23, 2010, http://www.russellnorthe.com.au/_blog/Media/post/coalition_to_tackle_labor's_smart_meter_stuff_up/

45. Kenneth Walsh, "Carol Browner on Climate Change: 'The Science Has Just Become Incredibly Clear,'" *US News*, Mar. 9, 2009, http://www.usnews.com/news/energy/articles/2009/03/09/on-climate-change-the-science-has-just-become-incredibly-clear

46. "The UK Low Carbon Transition Plan," July 20, 2009, http://centralcontent.fco.gov.uk/central-content/campaigns/act-on-copenhagen/resources/en/pdf/DECC-Low-Carbon-Transition-Plan

47. John Carey, "Obama's Smart-Grid Game Plan," *Business Week*, October 27, 2009, http://www.businessweek.com/technology/content/oct2009/tc20091027_594339.htm

48. US Green Building Council, http://www.usgbc.org/DisplayPage.aspx?CategoryID=19

49. William R. Prindle, "From Shop Floor to Top Floor: Best Business Practices in Energy Efficiency," ICF International, April 2010, http://www.pewclimate.org/energy-efficiency/corporate-energy-efficiency-report

50. Ibid

51. *Why We Disagree About Climate Change: Understanding Controversy, Inaction and Opportunity* by Mike Hulme, (Cambridge University Press, 2009), p. 341; Spy image by Setreset under Creative Commons Attribution License

52. "Climate Issues May Limit Flights," *Belfast Telegraph*, February 7, 2009, http://www.belfasttelegraph.co.uk/news/environment/climate-issues-may-limit-flights-14177669.html

53. "UK Air Passenger Duty," HM Revenue and Customs, http://www.hmrc.gov.uk/air-passenger-duty/index.htm

54. "Dutch Government Ditches Eco Ticket Tax in Efforts to Halt Declining Traffic at Amsterdam Schiphol," Green Air On Line, March 30, 2009, http://www.greenaironline.com/news.php?viewStory=412

55. "Heathrow Airport Third Runway Plans are Untenable, Says Judge," *Telegraph*, March 26, 2010, http://www.telegraph.co.uk/travel/travelnews/7527255/Heathrow-Airport-third-runway-plans-are-untenable-says-judge.html

56. Gordon Pirie, "Airport Agency: Globalization and (Peri)urbanism," Observatoire Reunionnais des Arts, http://laboratoires.univ-reunion.fr/oracle/documents/421.html

57. Zeppelin image by M. Trischler under Creative Commons License; Leaf image by uneekGrafix under GNU Public License

58. "Carbon Offset Calculator," http://www.mygreenflight.com/carbon-calculator

59. Michael Winter, "Congress kills funding for Obama's high-speed rail," *USA Today*, Nov. 17, 2011, http://content.usatoday.com/communities/ondeadline/post/2011/11/congress-kills-funding-for-obamas-high-speed-rail-plan/1

60. "High-Speed Intercity Passenger Rail Program, Dept. of Transportation, http://www.fra.dot.gov/rpd/passenger/2243.shtml

61. Peter Glover, "Europe's Doomed Flight of Decarbonizing Fancy," *Energy Tribune*, Jan. 10, 2012, http://www.energytribune.com/articles.cfm/9530/Europes-Doomed-Flight-of-Decarbonizing-Fancy

62. Andrew Parker, "Turbulence hits EU airline pollution scheme," *Financial Times*, Oct. 3, 2011, http://www.ft.com/intl/cms/s/0/cca023aa-edb3-11e0-a9a9-00144feab49a.html#axzz1rOAfkKVc

63. Glover (See no. 61)

64. PETA, http://www.peta.org/b/thepetafiles/archive/2007/08/29/PETA-to-Gore-Too-Chicken-to-Go-Veg.aspx

65. Sue Blaine, "COP-17: EU adamant on aviation emissions," *Business Day*, Dec. 8, 2011, http://www.businessday.co.za/articles/Content.aspx?id=160655

66. PETA (See no. 64)

67. "Get Back in the Car: Vegetarian IPCC Chairman Rajendra Pachauri Says Less Meat Will Slow Global Warming More," *Science 2.0*, September 6, 2008, http://www.science20.com/news_releases/get_back_in_the_car_vegetarian_ipcc_chairman_rajendra_ pachauri_says_less_meat_will_slow_global_warming_more

68. "Saving Carbon, Improving Health," National Health Service, January 2009, http://www.sdu.nhs.uk/documents/publications/1237308334_qylG_saving_carbon,_improving_health_nhs_carbon_reducti.pdf.

69. Martin Hickman, "McCartney Urges Meat-free Days to Tackle Climate Change," *Independent*, June 15, 2009, http://www.independent.co.uk/life-style/food-and-drink/news/mccartney-urges-meatfree-days-to-tackle-climate-change-1705289.html

70. Support Meat Free Monday, http://www.supportmfm.org/supporters/

71. Tristram Stuart, "Can Vegetarians Save the World?," *Guardian*, May 16, 2009, http://www.guardian.co.uk/environment/2009/may/16/ghent-belgium-vegetarian-town-environment

72. "Eat Kangaroo to Save the Planet" *BBC News*, August 9, 2009, http://news.bbc.co.uk/2/hi/uk_news/7551125.stm

73. "United Nations Calls for Insect Diet to Replace Red Meat," *MeatInfo*, August 2, 2010, http://www.meatinfo.co.uk/news/fullstory.php/aid/11121/United_Nations_calls_for_insect_diet_to_replace_red_meat.html

74. Insect photo by Meutia Chaerani/Indradi Soemardjan under GFDL; Comedy image by Booyabazook under GFDL

75. Louise Gray, "Fat People Causing Climate Change, Says Sir Jonathan Porritt," *Telegraph*, June 3, 2009, http://www.telegraph.co.uk/earth/environment/climatechange/5436335/Fat-people-causing-climate-change-says-Sir-Jonathan-Porritt.html

76. "Researchers Suggest Link Between Obesity & Global Warming," *Scotland Food and Drink*, Aug. 16, 2011, http://www.scotlandfoodanddrink.org/news/article-info/2617/scots-researchers-suggest-link-between-obesity-and-global-warming.aspx

77. *The Population Bomb* by Paul Ehrlich and David Brower (Sierra Club-Ballantine, 1970), p. 5

78. Paul Ehrlich speech at British Institute for Biology, September 1971; Crystal ball image by Eva Kröcher under GFDL

79. "World Population Prospects," UN Department of Economic and Social Affairs, http://esa.un.org/unpd/wpp2008/tab-sorting_fertility.htm

80. "World Fertility Report 2007," UN Department of Economic and Social Affairs, http://www.un.org/esa/population/publications/worldfertilityreport2007/wfr2007-text.pdf

81. Steve Connor, "Overpopulation is Main Threat to Planet," *Independent*, January 7, 2006, http://www.independent.co.uk/environment/overpopulation-is-main-threat-to-planet-521925.html

82. "Facing a Changing World: Women, Population and Climate," UN Population Fund, 2009, http://www.unfpa.org/swp/2009/en/

83. Brent Baker, "Turner: Global Warming Will Cause Mass Cannibalism, Insurgents Are Patriots," *Newsbusters*, April 2, 2008, http://newsbusters.org/blogs/brent-baker/2008/04/02/turner-iraqi-insurgents-patriots-inaction-warming-cannibalism

84. Patrick Goodenough, "Chinese Minister Links One-Child Policy to Emissions Reduction at Climate Conference," *CNSNews*, December 11, 2009, http://www.cnsnews.com/node/58412

85. Edwin Mora, "NYT Environment Reporter Floats Idea: Give Carbon Credits to Couples that Limit Themselves to One Child," *CNSNews*, October 19, 2009, http://www.cnsnews.com/news/article/55667

86. *The Skeptical Environmentalist: Measuring the Real State of the World* by Bjorn Lomborg (Cambridge University, 2001), p. 352

87. *Climate Change and the Failure of Democracy* by David Shearman and Joseph Smith (Praeger, 2007)

88. Leo Hickman, "James Lovelock on the Value of Sceptics and Why Copenhagen was Doomed," *Guardian*, March 29, 2010, http://www.guardian.co.uk/environment/blog/2010/mar/29/james-lovelock

89. Robert Newman, "It's Capitalism or a Habitable Planet - You Can't Have Both," *Guardian*, February 2, 2006, http://www.guardian.co.uk/environment/2006/feb/02/energy.comment

90. *Do the Right Thing: the People's Economist Speaks* by Walter E. Williams (Hoover, 1995), p. 105; Flag image by Ssolbergj under GFDL

91. Evo Morales Ayma, "Evo Morales on Climate Change: Save the Planet from Capitalism!," *Links International Journal of Socialist Renewal*, November 28, 2008, http://links.org.au/node/769

92. "Sustainable Lifestyles and Educations for Sustainable Consumption," United Nations Environmental Programme, http://esa.un.org/marrakechprocess/pdf/issues_sus_lifestyles.pdf

93. Timothy Wirth address at the 1992 Earth Summit in Rio de Janiero, http://www.climatedepot.com/a/9657/Prof-Larry-Bell-of-Forbes-fires-back-at-RealClimateorg-for-their-desperate-hit-and-run-tactics-against-skeptics; Spy image by Setreset under Creative Commons Attribution License

94. Louise Gray, "Cancun Climate Change Summit: Scientists Call for Rationing in Developed World," *Telegraph*, November 29, 2010, http://www.telegraph.co.uk/earth/environment/climatechange/8165769/Cancun-climate-change-summit-scientists-call-for-rationing-in-developed-world.html

95. Patrick Wintour, "UN Climate Change Deal Needs More Sacrifices by West, John Prescott Warns," *Guardian*, August 8, 2009, http://www.guardian.co.uk/environment/2009/aug/08/copenhagen-kyoto-climate-change-talks

96. *Climatism!* (See no. 10), p. 137-159

97. UN Secretary Ban Ki-Moon, remarks to the World Economic Forum, January 28, 2011, http://www.un.org/apps/news/infocus/sgspeeches/search_full.asp?statID=1057; Flag image by Ssolbergj under GFDL

98. Jacques Chirac, UNFCCC 6th Conference of the Parties speech, Nov. 20, 2000, http://sovereignty.net/center/chirac.html

99. Ban Ki-moon, "We Can Do It," *The New York Times*, Oct. 25, 2009, http://www.nytimes.com/2009/10/26/opinion/26iht-edban.html

100. Christian Wienberg, "Blizzard Dumps Snow on Copenhagen as Leaders Battle Warming," *Bloomberg*, Dec. 17, 2009, http://www.bloomberg.com/apps/news?pid=newsarchive&sid=a5wStc0K6jhY

101. Maurice Strong, "Stockholm to Rio: A Journey Down a Generation," http://sovereignty.net/p/gov/ggunreform.htm

102. Sunanda Creagh, "UN Meeting Moots WTO-Style Environment Agency," *Reuters*, February 26, 2010, http://www.reuters.com/article/2010/02/26/us-un-idUSTRE61P28920100226

103. Anders, "Czech President Klaus: Climate Alarmism is about World Governance and Freedom," *Euro-med*, Mar 17, 2008, http://euro-med.dk/?p=627

104. "Outcome of the work of the Ad Hoc Working Group on long-term Cooperative Action under the Convention," UNFCCC,

http://unfccc.int/files/meetings/cop_16/application/pdf/cop16_lca.pdf

105. Raven Clabough, "UN Official Admits Cap and Trade is Wealth Redistribution," *New American*, November 19, 2010, http://www.thenewamerican.com/index.php/world-mainmenu-26/europe-mainmenu-35/5253-un-official-admits-cap-and-trade-is-wealth-redistribution; Spy image by Setreset under Creative Commons Attribution License

106. Income and energy usage from "Key World Energy Statistics 2009, International Energy Agency, http://www.iea.org/textbase/nppdf/free/2009/key_stats_2009.pdf, vehicle usage from the German Federal Statistical Office (Destatis) country reports, http://www.destatis.de/jetspeed/portal/cms/Sites/destatis/Internet/EN/Navigation/Homepage__NT.psml

107. Low Carbon Development Path for Asia and the Pacific, December, 2010, http://www.unescap.org/esd/publications/energy/Series/2010/energy-publication-december-2010.pdf

108. Ibid

109. "Indoor Air Pollution and Health, World Health Organization, June, 2005, http://www.who.int/mediacentre/factsheets/fs292/en/index.html

110. Photograph by Zenman under GFDL

111. "World Bank Approves Multi-Billion-Dollar Loan for Coal-Fired Power Plant in South Africa," *Democracy Now*, April 9, 2010, http://www.democracynow.org/2010/4/9/world_bank_approves_multibillion_dollar_loan

112. Sarah van Schagen, "A New Reality Series Reveals What It's Like Living with Eco-Celeb Ed Begley Jr.," *Grist*, January 2, 2007, http://www.grist.org/article/begley/

113. Wind turbines in the Tehachapi Mountains of California, photograph by Stan Shebs under GFDL

114. *Climatism!* (See no. 10), p. 308-312

115. *Eco-Imperialism: Green Power, Black Death* by Paul Driessen (Free Enterprise Press, 2003), p. 42

116. Julia Gillard, "Carbon Price Now or We'll Pay Later," *Sydney Morning Herald National Times*, November 17, 2010, http://www.smh.com.au/opinion/politics/carbon-price-now-or-well-pay-later-20101116-17vti.html

117. St. Bernard photograph from Wikipedia under Creative Commons Generic License

118. Marc Morano, "Inhofe Slams New Cap-and-Trade Bill as All Economic Pain for No Climate Gain," US Senate Committee on Environment and Public Works: Minority Page, October 2007, http://epw.senate.gov/public/index.cfm?FuseAction=Minority.Blogs&ContentRecord_id=b4f81115-802a-23ad-4e54- f0137d7a406f&Issue_id=

119. Jeff Coelho, "Global carbon market value rises to record $176 billion," *Reuters*, May 30, 2012, http://www.reuters.com/article/2012/05/30/us-world-bank-carbon-idUSBRE84T08720120530

120. "Europe's Dirty Secret: Why the EU emissions Trading Scheme Isn't Working," Open Europe, August, 2007, http://www.openeurope.org.uk/research/etsp2.pdf

121. "About CDM," UNFCCC, http://cdm.unfccc.int/about/index.html

122. Paul Watson, "The Beginning of the End for Life as We Know it on Planet Earth," Sea Shepherd Deutschland, May 4, 2007, http://de.seashepherd.org/news-and-media/editorial-070504-1.html

123. Nils Klawitter, "Will Trading System Encourage Emissions?" *Spiegel Online*, December 30, 2010, http://www.spiegel.de /international /business/0,1518,736801,00.html

124. John Heilprin, "UN Carbon Trading Scheme: $2.7 Billion Market Could Be Biggest Environmental Scandal in History," *Huffington Post*, October 21, 2010, http://www.huffingtonpost.com/2010/08/23/un-carbon-trading-scheme-_n_690958.html

125. Ibid

126. "Denmark Rife with CO_2 Fraud," *The Copenhagen Post*, December 1, 2009, http://www.cphpost.dk/news/national/88-national/47643-denmark-rife-with-co2-fraud.html

127. Mark Shapiro, "Carbon Carousel: European Market a Haven for Tax Fraud," Center for Investigative Reporting, June 15, 2010, http://www.pbs.org/frontlineworld/stories/carbonwatch/2010/06/carbon-carousel-europes-market-haven-for-tax-fraud.html

128. Rowena Mason, "The Great Carbon Trading Scandal," *Telegraph*, January 30, 2011, http://www.telegraph.co.uk/finance/newsbysector/energy/8290533/The-great-carbon-trading-scandal.html

129. Alexander Jung, "The Pitfalls of Europe's New Emissions Trading System," *Spiegel*, December 30, 2010, http://www.spiegel.de/international/business/0,1518,736798,00.html

130. Ibid

131. "CRC - carbon reduction commitment," http://www.ukcrc.co.uk/

132. "Tata Steel fears effects of carbon tax in Budget, *BBC News*, Mar. 24, 2011, http://www.bbc.co.uk/news/uk-wales-12847929

133. Walter Olson, "Australians to punish carbon tax criticism," *US Action News*, http://usactionnews.com/2011/11/australians-to-punish-carbon-tax-criticism/

134. "Australian Cabinet to vote on carbon tax," *Energy Daily*, Aug. 17, 2011, http://www.energy-daily.com/reports/Australian_Cabinet_to_vote_on_carbon_tax_999.html

Chapter 3: The Simple Science of Man-made Global Warming

1. Climatic Research Unit, University of East Anglia, http://www.cru.uea.ac.uk/

2. Intergovernmental Panel on Climate Change, Third Assessment Report, Working Group I, Summary for Policy Makers, (2001) p. 2

3. CRU HadCRUT3 data set, http://www.metoffice.gov.uk/hadobs/hadcrut3/index.html
4. National Oceanic and Atmospheric Administration, Earth System Research Laboratory, http://www.esrl.noaa.gov/gmd/ccgg/trends/
5. Intergovernmental Panel on Climate Change, Fourth Assessment Report, Synthesis Report, (2007), p. 46
6. *Climate Change: Picturing the Science* by Gavin Schmidt and Joshua Wolfe (Norton, 2009), p. 31, 32, 81, 87, 96, 102, 108, 109
7. Mead Gruver, "Construction begins on $100m climate supercomputer in Wyo.," *USAToday*, June 15, 2010, http://www.usatoday.com/weather/research/2010-06-15-climate-supercomputer-wyoming_N.htm
8. Megan McArdle, "Scientists: Aliens May Punish Our Species for Climate Change," *Atlantic*, August, 2011, http://www.theatlantic.com/national/archive/2011/08/scientists-aliens-may-punish-our-species-for-climate-change/243886/
9. S. Fred Singer, "Nature, Not Human Activity, Rules the Climate," The Heartland Institute, 2008

Chapter 4: History Shows Global Warming *Not* Abnormal

1. The Nobel Peace Prize 2007, http://nobelprize.org/nobel_prizes/peace/laureates/2007/ipcc-lecture_en.html
2. Intergovernmental Panel on Climate Change, Fourth Assessment Report, Synthesis Report, (2007), p. 39
3. Intergovernmental Panel on Climate Change, Fourth Assessment Report, Working Group I, Chapter 6 (2007), p. 468
4. Clown photograph by Rick Dikeman under GFDL
5. The Official Tourism Site of Greenland, http://www.greenland.com/en/about-greenland/kultur-sjael/historie/vikingetiden/erik-den-roede.aspx
6. Bianca, "The Inuit of the Arctic Regions of Canada, Greenland, Russia & Alaska, http://atlantisonline.smfforfree2.com/index.php/topic,16044.25/wap2.html
7. "Cologne Cathedral," http://www.sacred-destinations.com/germany/cologne-cathedral
8. Photo by Jacob Munksgaard
9. Luc Sorel, "Climatic changes, the Norsemen and the Modern Man, 2001, http://www.lucsorel.com/media/research/2001_ClimaticChangesNorsemenAndTheModernMan_Sorel.pdf
10. Ibid
11. Fred Singer and Dennis Avery, "The Physical Evidence of Earth's Unstoppable 1,500-year Climate Cycle, NCPA Policy Report No. 279, September, 2005, http://www.ncpa.org/sub/dpd/index.php?Article_ID=2319
12. Image of painting by Thomas Wyke, 1683
13. Devin Powell, "Columbus blamed for Little Ice Age," *ScienceNews*, October 13, 2011, http://www.sciencenews.org/index.php/feed/label_id/2529/name/Paleontology.rss/view/generic/id/335168/title/Columbus_blamed_for_Little_Ice_Age
14. Haken Grudd, "Tornetrask tree-ring width and density AD 500-2004: a test of climatic sensitivity and a new 1500-year reconstruction of north Fennoscandian summers," *Climate Dynamics*, vol. 31, 30 January, 2008, pp.842-857
15. Karin Holmgren et al., "A preliminary 3000-year regional temperature reconstruction for South Africa," *South African Journal of Science*, Vol. 97, 2001, pp. 49-51
16. Grudd, (See no. 14), and *CO₂Science*, http://www.co2science.org/data/mwp/studies/l1_tornetrask.php
17. Holmgren, (See no. 15), and *CO₂Science*, http://www.co2science.org/data/mwp/studies/l1_coldaircave.php
18. Delia Oppo et al., "2,000-year-long temperature and hydrology reconstructions from the Indo-Pacific warm pool," *Nature*, Vol. 460, 2009, pp. 1113-1116
19. Ibid
20. Quansheng Ge et al., "Winter half-year temperature reconstruction for the middle and lower reaches of the Yellow River and Yangtze River, China, during the past 2000 years," *The Holocene*, Vol. 13, 2003, pp. 933-940
21. Andrei Andreev et al., "Environmental changes in the northern Altai during the last millennium documented in Lake Teletskoye pollen record," *Quaternary Research*, Vol. 67, 2007, pp. 394-399
22. Ge, (See no. 20), and *CO₂Science*, http://www.co2science.org/data/mwp/studies/l1_easternchina.php
23. Andreev, (See no. 21), and *CO₂Science*, http://www.co2science.org/data/mwp/studies/l1_laketeletskoye.php
24. Julie Richey et al., "1400 yr multiproxy record of climate variability from the northern Gulf of Mexico," *Geology*, Vol. 35, 2007, pp. 423-426
25. Ibid and *CO₂Science*, http://www.co2science.org/data/mwp/studies/l1_pigmybasin.php
26. *CO₂Science*, http://www.co2science.org/data/mwp/mwpp.php
27. *Unstoppable Global Warming Every 1500 Years* by S. Fred Singer and Dennis Avery, (Rowman & Littlefield, 2007), p. 1
28. W. Dansgaard et al., "North Atlantic Climate Oscillations Revealed by Deep Greenland Ice Cores," *Climate Processes and Climatic Sensitivity*, 1984, pp. 288-298, adapted by Dennis Avery, Presentation at the International Conference on Climate Change, Mar. 9, 2009
29. "NASA Research Finds Last Decade was Warmest on Record, 2009 One of Warmest Years," NASA press release January 21, 2010, http://www.nasa.gov/home/hqnews/2010/jan/HQ_10-017_Warmest_temps.html
30. Bob Carter, video "Climate Change - Is CO2 the Cause?," http://www.youtube.com/watch?v=FOLkze-9Gcl; Polar bear photograph by Alan Wilson licensed under Creative Commons Attribution-Share Alike 3.0 Unported License.
31. William Happer, "The Truth About Greenhouse Gases," *First Things*, May 21, 2011,http://www.firstthings.com/article/2011/05/the-truth-about-greenhouse-gases

32. Intergovernmental Panel on Climate Change, Third Assessment Report, Summary for Policy Makers, (2007), p. 1

33. Intergovernmental Panel on Climate Change, Third Assessment Report (2001), Working Group I, Chapter 2, Figure 2.20, p. 134

34. David Deming, testimony before the US Senate Committee on Environment and Public Works, Dec. 12, 2006, http://epw.senate.gov/hearing_statements.cfm?id=266543

35. Intergovernmental Panel on Climate Change, First Assessment Report (1990), Figure 7.1c, p. 202

36. Intergovernmental Panel on Climate Change, Third Assessment Report, Working Group I, Summary for Policy Makers, (2001), p. 3, http://www.grida.no/publications/other/ipcc_tar/

37. Intergovernmental Panel on Climate Change, Fourth Assessment Report, Working Group I, Technical Summary, (2007), p. 55, http://www.ipcc.ch/publications_and_data/ar4/wg1/en/contents.html

38. Stephen McIntyre and Ross McKitrick, "Hockey sticks, principle components, and spurious significance," *Geophysical Research Letters*, vol. 32, Oct. 14, 2004

39. Ibid

40. "Global Warming or Cooling," http://www.lowerwolfjaw.com/agw/quotes.htm; Image by Eva Kröcher under GFDL

41. Adapted from *Climate4You*, http://www.climate4you.com/; data from NOAA Mauna Loa and from CRU HadCRUT3

42. E-mail from Kevin Trenberth to Myles Allen, Keith Briffa, James Hansen, Phil Jones, Thomas Karl, Michael Mann, Michael Oppenheimer, Benjamin Santer, Gavin Schmidt, Steven Schneider, Peter Stott, and Tom Wigley, "The Climategate Emails," Lavoisier Group, Oct. 12, 2009, http://www.lavoisier.com.au/articles/greenhouse-science/climate-change/climategate-emails.pdf

43. El Niño, National Oceanic and Atmospheric Adminstration, http://www.elnino.noaa.gov/

44. Adapted from *Climate4You*, http://www.climate4you.com/; data from University of Washington, JISAO

45. "The Pacific Decadal Oscillation," University of Washington, http://jisao.washington.edu/pdo/

46. Syun-Ichi Akasofu, "Is the Earth Still Recovering from the 'Little Ice Age?,' A Possible Cause of Global Warming," http://www.wright.edu/~guy.vandegrift/climateblog/s06/akasofu.LIAge.pdf

Chapter 5: Climate Science—The Rest of the Story

1. *Storms of My Grandchildren: The Truth About the Coming Climate Catastrophe and Our Last Chance to Save Humanity* by James Hansen (Bloomsbury USA, 2009), p. 49

2. After diagram by Timothy Ball; sun image from Solar and Heliospheric Observatory, NASA, http://sohowww.nascom.nasa.gov/; Jupiter and Io image from Don Davis; Cloud image by Przemyslaw under Creative Commons Attribution-Share Alike Generic License; Wave image by PDphoto; Coral reef image by Habeeb; Abissal fish image by Francesco Costa under GFDL; Volcano image by Austin Post; City image by Storylanding under Creative Commons License; Joshua tree image by Joho345 under Creative Commons License; Forest image by Malene Thyssen under GFDL; Glacier image by Tobias Alt under GFDL

3. Glenn Elert, "Mass of the Atmosphere," *The Physics Fact Book*, 1999, http://hypertextbook.com/facts/1999/LouiseLiu.shtml

4. Steve Connor, "Letters to a heretic: An E-mail conversation with climate change sceptic Professor Freeman Dyson," *Independent*, Feb. 25, 2011, http://www.independent.co.uk/environment/climate-change/letters-to-a-heretic-an-email-conversation-with-climate-change-sceptic-professor-freeman-dyson-2224912.html

5. Intergovernmental Panel on Climate Change, Fourth Assessment Report, Working Group I, Summary for Policymakers (2007), p. 3

6. Ibid, p. 2

7. Siple Dome drilling rig photograph from Los Alamos National Laboratory, http://public.lanl.gov/sprice/images/wais99/page2.html

8. Ice core photograph by Emily Stone, National Science Foundation, http://photolibrary.usap.gov/AntarcticaLibrary/CORECLOSEUP.JPG

9. Zabigniew Jaworowski, "Ice Core Data Show No Carbon Dioxide Increase," *21st Century*, Spring 1997, http://www.21stcenturysciencetech.com/2006_articles/IceCoreSprg97.pdf

10. Ibid

11. Ibid

12. Spiderwort photograph by David Byres, http://www1.fccj.cc.fl.us/dbyres/bsc_2010c.htm

13. Lenny Kouwenberg et al., "Application of Conifer Needles in the Reconstruction of Holocene CO_2, Levels," PhD Thesis, University of Utrecht, Jan. 16, 2004, http://igitur-archive.library.uu.nl/dissertations/2004-0128-122010/inhoud.htm

14. Thomas van Hoof et al., "Atmospheric CO_2 during the 13th Century AD: reconciliation of data from ice core measurements and stomatal frequency analysis," *Tellus*, 2005, 57B, pp. 351-355, http://www.phys.uu.nl/~wal/research/papers/hoofetal2005.pdf

15. Kouwenberg (See no. 13)

16. J. R. Petit et al., "Climate and Atmospheric History of the Past 420,000 Years from the Vostok Ice Core, Antarctica," *Nature*, v. 399, June 3, 1999, pp. 429-436

17. The 7-12°C according to *Unstoppable Global Warming Every 1500 Years* by S. Fred Singer and Dennis Avery, but scientists are not really searching for SUV and coal plant remains; Comedy image by Booyabazook under GFDL

18. C. Lorius et al., "The Ice-Core Record: Climate Sensitivity and Future Greenhouse Warming, "*Nature*, v. 347, Sep. 13,

1990, pp. 139-145

19. *Canon of Insolation and the Ice-Age Problem* by Milutin Milankovic, 1941, (Agency for Textbooks, 1998)

20. Hubertus Fischer et al., "Ice Core Records of Atmospheric CO_2 around the Last Three Glacial Terminations," *Science*, v. 283, mar. 12, 1999, pp. 1712-1714

21. Intergovernmental Panel on Climate Change, Fourth Assessment Report, Working Group I, (2007), p. 501

22. Ibid, p. 515

23. Ibid, p. 515

24. Tom Segelstad, "Carbon Cycle Modeling and the Residence Time of Natural and Anthropogenic Atmospheric CO_2: On the Constructon of the 'Greenhouse Effect Global Warming Dogma," http://www.geocraft.com/WVFossils/Reference_Docs/Carbon_cycle_update_Segalstad.pdf

25. "Henry's Law and the Solubility of Gases," Introductory University Chemistry I, http://dwb.unl.edu/teacher/nsf/c09/c09links/www.chem.ualberta.ca/courses/plambeck/p101/p01182.htm

26. "Climate change' could spark more volcanoes, earthquakes and tsunamis,'" *Daily Mail*, Apr. 19, 2010, http://www.dailymail.co.uk/sciencetech/article-1267137/Climate-change-spark-volcanoes-earthquakes-tsunamis.html

27. Barry Woods, "Simple Physics - In reality my feather blew up into a tree," *WattsUpWithThat*, Dec. 28, 2010, http://wattsupwiththat.com/2010/12/28/simple-physics-in-reality-my-feather-blew-up-into-a-tree/

28. Gavin Schmidt et al., "The attribution of the present-day total greenhouse effect," NASA Goddard Institute for Space Studies, draft paper, August 10, 2010

29. It is true that water vapor is nature's most abundant greenhouse gas, but it's not true that the EPA is considering whether to declare water a pollutant; Comedy image by Booyabazook under GFDL

30. Roy Spencer, "Global Warming and Nature's Thermostat," http://petesplace-peter.blogspot.com/2008/01/global-warming-and-natures-thermostat.html

31. Richard Lindzen, "Understanding Common Climate Claims," proceedings of the 2005 ERICE Meeting of the World Federation of Scientists on Global Emergencies, 2005

32. David Archibald, "Solar Cycle 24: Implications for the United States," presented at the International Conference on Climate Change, March, 2008

33. Syukuro Manabe and Richard Wetherald, "Thermal Equilibrium of the Atmosphere with a Given Distribution of Relative Humidity," *Geophysical Fluid Dynamics Laboratory*, ESSA, pp.241-259, Nov. 2, 1996

34. Intergovernmental Panel on Climate Change, Fourth Assessment Report, Working Group I, Summary for Policy Makers (2007), p. 12

35. Adapted from *Climate4You*, http://www.climate4you.com/, data from International Cloud Climatology Project

36. Richard Lindzen and Yong-Sang Choi, "On the Observational Determination of Climate Sensitivity and Its Implications," *Asia-Pacific Journal of Atmospheric Science*, 47(4), pp. 377-390, May 22, 2011

37. Roy Spencer and William Braswell, "On the Diagnosis of Radiative Feedback in the Presence of Unknown Radiative Forcing," *Journal of Geophysical Research*, v. 115, August 24, 2010

38. David Douglass et al., "A comparison of tropical temperature trends with model predictions," *International Journal of Climatology*, October 11, 2007

39. David Evans, "Climate models go cold," *Financial Post*, Apr. 7, 2011, http://opinion.financialpost.com/2011/04/07/climate-models-go-cold/

40. Andrew Orlowski, "Would putting all the climate scientists in a room solve global warming...," *The Register,* May 13, 2011, http://www.theregister.co.uk/2011/05/13/downing_cambridge_climate_conference/print.html

41. Intergovernmental Panel on Climate Change, Fourth Assessment Report, Working Group I, (2007), p. 675

42. "Temperature Trends in the Lower Atmosphere," US Climate Change Science Program, April 2006, p. 116

43. Robert Allen and Steven Sherwood, "Utility of Radiosonde Wind Data in Representing Climatological Variations of Trospheric Temperature and Baroclinicity in the Western Tropical Pacific," *Journal of Climate*, v. 20, Feb. 27, 2007

44. Intergovernmental Panel on Climate Change, Fourth Assessment Report, Working Group I, (2007), p. 691

45. J. Beer et al., "The role of the sun in climate forcing," *Quaternary Science Reviews*, 19, 2000, pp. 403-415

46. Douglas Hoyt and Kenneth Schatten, "Group Sunspot Numbers: A New Solar Activity Reconstruction," *Solar Physics*, v. 181, iss. 2, pp. 491-512 (1998)

47. Douglas Hoyt and Kenneth Schatten, "A Discussion of Plausible Solar Irradiance Variations, 1700-1992," *Journal of Geophysical Research*, v. 98, no. A11, pp. 18,895-18,906, Nov. 1, 1993

48. Ibid

49. Hoyt (See no. 46)

50. Diagram adapted from Climate Data Information, http://www.climatedata.info/Forcing/Forcing/sunspots.html

51. Solar and Heliospheric Observatory, NASA, http://sohowww.nascom.nasa.gov/

52. US Air Force photograph, 2006

53. Henrik Svensmark, "Cosmoclimatology: a new theory emerges," *Astronomy and Geophysics*, Feb. 2007, v. 48

54. "Scientists Trace Heat Wave to Massive Star at Center of Solar System," *The Onion*, August 8, 2011, http://www.theonion.com/articles/scientists-trace-heat-wave-to-massive-star-at-cent,21088/; Comedy image by Booyabazook under GFDL

55. Nigel Marsh and Henrik Svensmark, "Cosmic Rays, Clouds, and Climate," *Space Science Reviews*, v. 94, pp. 215-230, May 10, 2000

56.   "The SKY Experiment," http://www.space.dtu.dk/English/Research/Research_divisions/Sun_Climate/Experiments_SC/SKY.aspx

57.   Nigel Calder, "CERN experiment confirms cosmic ray action," Aug. 24, 2011, http://calderup.wordpress.com/2011/08/24/cern-experiment-confirms-cosmic-ray-action/

58.   *The Chilling Stars: A New Theory of Climate Change* by Henrik Svensmark and Nigel Calder, (Totem Books, 2003)

59.   "Scientists at Aarhus University (AU) and the National Space Institute (DTU Space) Show that Particles from Space Creates Cloud Cover," Aarhus University, May 16, 2011, http://science.au.dk/en/news-and-events/news-article/artikel/forskere-fra-au-og-dtu-viser-at-partikler-fra-rummet-skaber-skydaekke/

60.   "CERN's CLOUD experiment provides unprecedented insight into cloud formation," CERN press release, August 25, 2011, http://press.web.cern.ch/press/PressReleases/Releases2011/PR15.11E.html

61.   Image by Joanne Nova, 2011, http://joannenova.com.au/

62.   Andrew Orlowski, "CERN 'gags'physicists in cosmic ray climate experiment," *The Register*, July 18, 2011, http://www.theregister.co.uk/2011/07/18/cern_cosmic_ray_gag/

63.   Henrik Svensmark, "Cosmic Rays and Earth's Climate," Aug. 13, 1999, http://ruby.fgcu.edu/courses/twimberley/EnviroPhilo/Svensmark.pdf

Chapter 6: The Oceans are Rising! The Oceans are Rising!

1.   "Climate Change and Sea Level," Department of Geosciences Environmental Studies Laboratory, Univ. of Arizona, 2011, http://www.geo.arizona.edu/dgesl/research/other/climate_change_and_sea_level/climate_change_and_sea_level.htm

2.   *The Economics of Climate Change: The Stern Review* by Nicholas Stern (Cambridge University Press, 2007)

3.   *An Inconvenient Truth: The Planetary Emergency of Global Warming and What We Can Do About It* by Al Gore (Rodale, 2006)

4.   Jim Hansen, "Climate Change: On the Edge," *Independent*, February 17, 2006, http://www.independent.co.uk/environment/climate-change-on-the-edge-466818.html

5.   Jonathan Amos, "Arctic summers ice-free 'by 2013,'" *BBC News*, Dec. 12, 2007, http://news.bbc.co.uk/2/hi/7139797.stm

6.   Ibid

7.   Andrew Revkin, "Arctic Melt Unnerves the Experts," *New York Times*, Oct. 2, 2007, http://www.nytimes.com/2007/10/02/science/earth/02arct.html

8.   "The Cryosphere Today," The Polar Research Group, Department of Atmospheric Sciences, University of Illinois, http://arctic.atmos.uiuc.edu/cryosphere/

9.   Anthony Watts, "You ask, I provide. November 2nd, 1922. Arctic Ocean Getting Warm; Seals Vanish and Icebergs Melt," *WattsUpWithThat*, March 16, 2008, http://wattsupwiththat.com/2008/03/16/you-ask-i-provide-november-2nd-1922-arctic-ocean-getting-warm-seals-vanish-and-icebergs-melt/

10.   Cryosphere (See no. 8)

11.   Adapted from *Climate4You*, http://www.climate4you.com/, data from John Christy and Roy Spencer, University of Alabama Huntsville, 2011

12.   British Antarctic Survey, http://www.nerc-bas.ac.uk/icd/gjma/temps.html

13.   From Steve Deyo diagram, copyright University Corporation for Atmospheric Research, 2008, http://www.fin.ucar.edu/netpub/server.np?find&site=imagelibrary&catalog=catalog&template=detail.np&field=itemid&op=matches&value=2670

14.   Amundsen-Scott South Pole station, US government photo

15.   "Amundsen-Scott South Pole Station," National Science Foundation Office of Polar Programs, http://www.nsf.gov/od/opp/support/southp.jsp

16.   Greenland ice melt photo by Roger Braithwaite, University of Manchester, NASA website, http://www.giss.nasa.gov/research/briefs/gornitz_09/

17.   GISS Surface Temperature Analysis, NASA Goddard Institute for Space Studies, http://data.giss.nasa.gov/cgi-bin/gistemp/findstation.py?datatype=gistemp&data_set=1&name=&world_map.x=285&world_map.y=55

18.   Karen Jensen, "'Glacier Girl' The Back Story," *Air & Space Magazine*, July 1, 2007, http://www.airspacemag.com/history-of-flight/FEATURE-glaciergirl-backstory.html

19.   Jeffrey Kluger, "How the Ice in Your Drink is Imperiling the Planet," *Time.com*, Apr. 14, 2011, http://ecocentric.blogs.time.com/2011/04/14/how-the-ice-in-your-drink-is-imperiling-the-planet/

20.   Jensen (See no. 18)

21.   Adapted from NASA, 2007, http://www.giss.nasa.gov/research/briefs/gornitz_09/

22.   Arthur Robinson et al., "Environmental Effects of Increased Atmospheric Carbon Dioxide," *Journal of American Physicians and Surgeons*, v. 12, pp. 79-90, 2007

23.   "From under water, Maldives sends warning on climate change," *CNN World*, Oct. 17, 2009, http://articles.cnn.com/2009-10-17/world/maldives.underwater.meeting_1_maldives-climate-change-sea-levels?_s=PM:WORLD

24.   Robert Engelman, "Carl Sagan: Fossil Fuels Bring Trouble," *The Vindicator*, Dec. 12, 1985, http://news.google.com/newspapers?id=rKM_AAAAIBAJ&sjid=RVYMAAAAIBAJ&pg=3633,6912687&dq=global+warming&hl=en

25.   Broward County Climate Action Plan, May 2010

26.   Ibid

27. John Upton, "Bay Area Adopts Historic Climate-Change Rules, *The Bay Citizen*, Oct. 6, 2011, http://www.baycitizen.org/development/story/bay-area-adopts-sea-level-rise-building/

28. George Quraishi, "Frostbite Chomps Arctic Ocean 2007 Expedition," *National Geographic Adventure Magazine*, Apr. 13, 2007, http://www.nationalgeographic.com/adventure/news/liv-arnesen-frostbite.html

29. Andrew Bolt, "Another polar rescue must send chills down the spines of alarmists," *Herald Sun*, Apr. 23, 2010 http://www.heraldsun.com.au/opinion/another-polar-rescue-must-send/story-e6frfhqf-1225856131380

30. Stefan Lovgren, "Warming to Cause Catastrophic Rise in Sea Level," *National Geographic News*, Apr. 26, 2004, http://news.nationalgeographic.com/news/2004/04/0420_040420_earthday.html

31. Bolt (See no. 29)

32. Chris Folland, presentation to climatologists, August 13, 1991, *Sound and Fury: The Science and Politics of Global Warming* by Patrick Michaels (Cato Institute, 1992), p. 83

33. Willie Soon, Presentation at the International Conference on Climate Change, June 2, 2009

Chapter 7: Wild Weather and Snow Follies

1. Tony Hake, "Five killed, dozens injured when winds topple Indiana State Fair stage," *Examiner*, August 15, 2011, http://www.examiner.com/natural-disasters-in-national/five-killed-dozens-injured-when-winds-topple-indiana-state-fair-stage

2. Scott Whitlock, "Diane Sawyer Uses Wind Disaster to Hype Global Warming: 'Weather Gone Wild,'" *Newsbusters*, August 16, 2011, http://newsbusters.org/blogs/scott-whitlock/2011/08/16/diane-sawyer-uses-wind-disaster-hype-global-warming-weather-gone-wild

3. 'Weather and Climate in Singapore," *Guide Me Singapore*, http://www.guidemesingapore.com/relocation/introduction/climate-in-singapore

4. "Meet Tom Udall: New Mexico Senator is Determined to Warn America about Climate Change," http://climateadaptation.tumblr.com/post/5067881286/tom-stewart-udall-utah-climate-change

5. Bill McKibben, "Global Warming's Heavy Cost," *The Daily Beast*, August 25, 2011, http://www.thedailybeast.com/articles/2011/08/25/hurricane-irene-can-be-tied-to-global-warming-says-bill-mckibben.html

6. Sharon Begley, "Are you ready for more?," *Newsweek*, May 29, 2011, http://www.thedailybeast.comnewsweek/2011/05/29/are-you-ready-for-more.html

7. Ibid.

8. Alec Rawls, "Yes, impossibly stupid 'weather panic' is the new normal,'" *Watts Up With That?*, June 3, 2011, http://wattsupwiththat.com/2011/06/03/yes-impossibly-stupid-weather-panic-is-the-new-normal/

9. Joseph D'Aleo, "Uh, oh…the clash of ice and warmth brings storms," *Watts Up With That?*, April 9, 2011, http://wattsupwiththat.com/2011/04/09/uh-oh-2/

10. Kerry Emanuel, "Increasing destructiveness of tropical cyclones over the past 30 years," *Nature*, vol. 436, August 4, 2005, pp. 686-688

11. William Gray, "Gross Errors in the IPCC-AR4 Report Regarding Past and Future Changes in Global Tropical Cyclone Activity," Science and Public Policy Institute, Oct. 11, 2011, http://scienceandpublicpolicy.org/originals/gross_errors_ipcc_ar4.html

12. Rawls (See no. 8)

13. Intergovernmental Panel on Climate Change, Fourth Assessment Report, Summary for Policy Makers, p. 15

14. Christopher Landsea et al., "Can We Detect Trends in Extreme Tropical Cyclones?," *Science*, vol. 313, July 28, 2006, pp. 452-454

15. *Hell and High Water: Global Warming—the Solution and the Politics—and What We Should Do* by Joseph Romm (HarperCollins, 2007), p. 90

16. Open Letter to the Community from Chris Landsea, January 17, 2005, http://www.climatechangefacts.info/ClimateChangeDocuments/LandseaResignationLetterFromIPCC.htm

17. Comedy image by Booyabazook under GFDL

18. Ryan Maue, Global Tropical Cyclone Activity Update, 2011, http://policlimate.com/tropical/

19. Gray (See no. 11)

20. "Satellite Studies," Center for Ocean-Atmospheric Prediction Studies, Florida State University, http://coaps.fsu.edu/satellite.shtml

21. Ryan Maue, "Global hurricane activity at historical lows," June 27, 2011, http://www.outlookseries.com/A0996/Science/3925_Ryan_Maue_FSU_Global_hurricane_activity_historical_lows_Ryan_Maue.htm

22. Jeff Masters, "Tropical Storm Paula Forming," Wunderblog, October 11, 2010, http://weather.wb11.com/blog/JeffMasters/comment.html?entrynum=1652&tstamp=&page=14

23. US Tornado Climatology, National Climatic Data Center, http://lwf.ncdc.noaa.gov/oa/climate/severeweather/tornadoes.html

24. Roy Spencer, "MORE Tornadoes from Global Warming? That's a Joke, Right?," April 29, 2011, http://www.drroyspencer.com/2011/04/more-tornadoes-from-global-warming-thats-a-joke-right/

25. Richard Lindzen, "Understanding Common Climate Claims," draft paper, 2005, http://www.geocraft.com/WVFossils/Reference_Docs/Lindzen_2005_Climate_Claims.pdf

26. *The Climate Crisis: An Introductory Guide to Climate Change* by David Archer and Stefan Rahmsdorf, (Cambridge

University Press, 2010), p. 147

27. Charles Onians, "Snowfalls are now just a thing of the past," *Independent*, March 20, 2000, http://www.independent.co.uk/environment/snowfalls-are-now-just-a-thing-of-the-past-724017.html

28. Ian Herbert, "Snowdon will be snow-free in 13 years, scientists warn, "*Independent*, January 18, 2007, http://www.independent.co.uk/environment/climate-change/snowdon-will-be-snowfree-in-13-years-scientists-warn-432596.html

29. T.P. Barnett et al., "Anthropogenic Global Warming and the Western USA's Diminishiing Snowpack: A Link," Waterwired, January 4, 2008, http://aquadoc.typepad.com/waterwired/2008/01/global-warming.html

30. *Climate Change: Picturing the Science* by Gavin Schmidt and Joshua Wolfe (Norton, 2009)

31. Benny Peiser, "Warm Bias: How the Met Office Misled the British Public," *gwpf.org*, December 18, 2010, http://www.thegwpf.org/uk-news/2073-warm-bias-how-the-met-office-mislead-the-british-public.html

32. Auslan Cramb, "Scottish ski industry could disappear due to global warming, warns Met Office," *Telegraph*, February 10, 2009, http://www.telegraph.co.uk/news/uknews/scotland/4579829/Scottish-ski-industry-could-disappear-due-to-global-warming-warns-Met-Office.html

33. Laura Clout, "British UFO sightings at 'bizarre' levels," *Telegraph*, July 7, 2008, http://www.telegraph.co.uk/news/newstopics/howaboutthat/2261941/British-UFO-sightings-at-bizarre-levels.html, Hot earth image by Jack1 under GFDL

34. Jonathan Harwood, "Climate sceptics seize on Scotland's coldest winter," *FirstPost*, February 2, 2010, http://www.thefirstpost.co.uk/59215,news-comment,news-politics,scotland-endures-coldest-winter-says-met-office

35. Peiser (See no. 31)

36. Frozen Britain seen from above, *BBC News*, January 7, 2010, http://news.bbc.co.uk/2/hi/8447023.stm

37. "Louise Ellman Asked to probe Met Office's 'Conflicting' Winter Weather Advice," *Local Transport Today*, January 15, 2011, http://www.thegwpf.org/uk-news/2267-ellman-asked-to-probe-met-offices-conflicting-winter-weather-advice.html

38. James Delingpole, "The Met office: lousier than a dead octopus," *Telegraph*, December 2, 2010, http://blogs.telegraph.co.uk/news/jamesdelingpole/100066366/why-did-we-slide-into-chaos-well-duh/

39. Paul Hudson, "December 2010 update: Second coldest since 1659," *BBC*, January 4, 2011, http://www.bbc.co.uk/blogs/paulhudson/2011/01/december-2010-update-second-co.shtml

40. "Heathrow airport triples snow clearance fleet," *BBCNews*, September 29, 2011, http://www.bbc.co.uk/news/uk-england-london-15105627

41. Anthony Watts, "There's no business like snow business," *WattsUpWithThat?*, February 23, 2010, http://wattsupwiththat.com/2010/02/23/theres-no-business-like-snow-business/

42. "United States of Snow," *CBSNews*, February 12, 2010, http://www.cbsnews.com/stories/2010/02/12/national/main6202529.shtml

43. "Winter of 2010-2011 in Europe," *Wikipedia*, http://en.wikipedia.org/wiki/Winter_of_2010%E2%80%932011_in_Europe

44. Anthony Watts, "Western snow pack is well above normal, Squaw Valley sets new all time snow record," *WattsUpWithThat?*, http://wattsupwiththat.com/2011/05/18/western-snow-pack-is-well-above-normal-squaw-valley-sets-new-all-time-snow-record/

45. Lake Powell Water Database, September 28, 2011, http://lakepowell.water-data.com/

46. Richard Alleyne, "Snow is consistent with global warming, say scientists," *Telegraph*, February 3, 2009, http://www.telegraph.co.uk/topics/weather/4436934/Snow-is-consistent-with-global-warming-say-scientists.html

47. Daniel Huber and Jay Gulledge, "Extreme weather and climate change: Understanding the link, managing the risk," Pew Center on Global Climate Change, June 2011, http://www.pewclimate.org/docUploads/white-paper-extreme-weather-climate-change-understanding-link-managing-risk.pdf

48. Photo by dbking under Creative Commons Attribution 2.0 License, Hot earth image by Jack1 under GFDL

49. Al Gore, "An answer for Bill," Al Journal, February 1, 2011, http://blog.algore.com/2011/02/an_answer_for_bill.html

50. James Delingpole, "Signs that show Man-made Global Warming is Definitely Still Happening," *Telegraph*, Dec. 3, 2010, http://blogs.telegraph.co.uk/news/jamesdelingpole/100066594/signs-that-show-man-made-global-warming-is-definitely-still-happening/

51. George Monbiot, "Cancún climate change summit: Is God determined to prevent a deal?," *Guardian*, December 2, 2010, http://www.guardian.co.uk/environment/georgemonbiot/2010/dec/02/cancun-climate-change-summit-monbiot

52. George Monbiot, "I'm all for putting more vehicles on our roads. As long as they're coaches," *Guardian*, Dec. 5, 2006, http://www.cicle.org/wordpress/2006/12/05/im-all-for-putting-more-vehicles-on-our-roads-as-long-as-theyre-coaches-2/

53. Anthony Watts, "'Gore effect' strikes Cancun Climate Conference 3 days in a row," *WattsUpWithThat?*, http://wattsupwiththat.com/2010/12/08/gore-effect-strikes-cancun-climate-conference-3-days-in-a-row/

54. Conor Clarke, "An Interview with Thomas Schelling, Part Two" *Atlantic Wire*, July 13, 2009, http://www.theatlantic.com/politics/archive/2009/07/an-interview-with-thomas-schelling-part-two/21273/

55. Frances Welch, "Moral outlook: earthquake, wind and fire," *Sunday Telegraph*, Oct. 9, 1995, http://john-adams.co.uk/wp-content/uploads/2010/02/houghton-and-god.pdf

56. Foresight, "International Dimensions of Climate Change," The Government Office for Science, London, 2011

57. Emily Oster, "Witchcraft, Weather and Economic Growth in Renaissance Europe," *Journal of Economic Perspectives*, vol. 18, no. 1, Winter 2004, pp. 215-228

Chapter 8: Big Whoppers About Climate Change

1. "WWF Calls on President, Congress for Leadership in Passing Climate and Energy Legislation," World Wildlife Fund, May 12, 2010, http://www.worldwildlife.org/who/media/press/2010/WWFPresitem16252.html
2. "Obama Advisor Carol Browner Addresses Green Mountain College Class of 2010," *PRNewswire*, May 15, 2010, http://www.prnewswire.com/news-releases/obama-advisor-carol-browner-addresses-green-mountain-college-class-of-2010-93852479.html
3. Ron Bailey, presentation at the Foundation for the Future, *Reason Magazine*, August 12, 2001, http://reason.com/archives/2001/08/12/state-of-the-world/singlepage
4. Donald Brown, "A New Kind of Crime Against Humanity?: The Fossil Fuel Industry's Disinformation Campaign On Climate Change," Oct. 24, 2010, http://rockblogs.psu.edu/climate/2010/10/a-new-kind-of-vicious-crime-against-humanity-the-fossil-fuel-industrys-disinformation-campaign-on-cl.html
5. Photo by Alan Murray-Rust under Creative Commons License
6. US Environmental Protection Agency, 2011, http://www.epa.gov/air/airpollutants.html
7. Ibid, "2011 Statistical Abstract," US Census Bureau, 2011, http://www.census.gov/compendia/statab/
8. Michael Schirber, "The Chemistry of Life: The Human Body," *Live Science*, Apr. 16, 2009, http://www.livescience.com/3505-chemistry-life-human-body.html
9. http://www.co2isgreen.org/default.aspx/MenuItemID/137/MenuGroup/Home.htm
10. http://www.co2science.org/education/experiments/center_exp/experiment1/exp1_home.php
11. Arthur Robinson et al., "Environmental Effects of Increased Atmospheric Carbon Dioxide," *Journal of American Physicians and Surgeons*, v. 12, pp. 79-90, 2007
12. Bread photo by David Monniaux under GFDL
13. William Happer, "The Truth About Greenhouse Gases," *First Things*, June/July 2011, http://www.firstthings.com/article/2011/05/the-truth-about-greenhouse-gases
14. Joanne Nova, radio interview with Mark Gillar, April 15, 2010, http://www.blogtalkradio.com/markgillar/2010/04/15/joann-nova-discusses-global-warming
15. "Full text: Blair's climate change speech," *Guardian*, Sep. 15, 2004, http://www.guardian.co.uk/politics/2004/sep/15/greenpolitics.uk
16. Edward Malbach, "The Health Community Should Reframe Climate Change as a Human Health Issue," *Cornerstone*, George Mason University, May 9, 2011, http://cornerstone.gmu.edu/articles/2927
17. Thomas Karl et al., "Global Climate Change Impacts in the United States," 2009, http://www.globalchange.gov/what-we-do/assessment/previous-assessments/global-climate-change-impacts-in-the-us-2009
18. Photograph by Norman Rogers, 2010
19. W. R. Keating et al., "Heat Related Mortality in Warm and Cold Regions of Europe: Observational Study," *British Medical Journal*, v. 321, Sep. 16, 2000, pp. 670-673
20. Matthew Falagas et al., "Seasonality of Mortality: The September Phenomenon in Mediterranean Countries," *Canadian Medical Association Journal*, Oct. 13, 2009
21. Ibid
22. Bjorn Lomborg, "Global Warming Will Save Millions of Lives," *Telegraph*, Mar. 13, 2009, http://www.telegraph.co.uk/comment/personal-view/4981028/Global-warming-will-save-millions-of-lives.html
23. "Global Warming Link to Kidney Stones," *The Times of India*, May 15, 2008
24. Karl (See no. 17)
25. Karl (See no. 17)
26. Andrew Restuccia, "House Dem: Climate Change Bigger Health Threat Than AIDS, Malaria," *The Hill*, Apr. 6, 2011, http://thehill.com/blogs/e2-wire/e2-wire/154251-house-dem-climate-change-bigger-health-threat-than-aids-malaria
27. "Flesh-Eating Disease Is On The Rise Due To Global Warming, Experts Warn," *ScienceDaily*, Aug. 16, 2007, http://www.sciencedaily.com/releases/2007/08/070815152912.htm; Hot earth image by Jack1 under GFDL
28. Karl (See no. 17)
29. Paul Reiter, "Kyoto Protocol is irrelevant to the spread of disease says expert," International Policy Network, Dec. 11, 2010, http://www.policynetwork.net/es/environment/media/kyoto-protocol-irrelevant-spread-disease-says-expert
30. Letter to President Obama and the Congress, Sep. 28, 2010, http://www.usclimatenetwork.org/resource-database/ph-groups-sign-on-climate-change-9.28.10
31. Eryka Bergey, "The Philadelphia Zoo unveils new exhibit made entriely of LEGO blocks," *Examiner.com*, August 11, 2010, http://www.examiner.com/life-photos-in-philadelphia/the-philadelphia-zoo-unveils-new-exhibit-made-entirely-of-lego-blocks
32. Mitch Taylor, presentation at the International Conference on Climate Change, Mar. 10, 2009
33. Clifford Kraus, "Bear Hunting Caught in Global Warming Debate," *New York Times*, May 27, 2006, http://www.nytimes.com/2006/05/27/world/americas/27bears.html?pagewanted=all
34. Comedy image by Booyabazook under GFDL
35. "Will Polar Bears Survive?" IUCN Polar Bear Specialist Group, July 6, 2009, http://www.polarbearsinternational.org/polar-bears/will-polar-bears-survive
36. "Cancun climate change summit: climate change killing polar bears,"*Telegraph*, Dec. 9, 2010, http://www.telegraph.co.uk/

earth/environment/climatechange/8190922/Cancun-climate-change-summit-climate-change-killing-polar-bears.html

37. Ian Sterling et al., "Polar Bear Population Status in the Northern Beaufort Sea," USGS Administrative Report, 2007

38. Kirk Meyers, "Canada's growing polar bear population 'becoming a problem,' locals say," *Examiner.com*, January 8, 2010, http://www.examiner.com/seminole-county-environmental-news-in-orlando/canada-s-growing-polar-bear-population-becoming-a-problem-locals-say

39. Joseph Abrams, "Out With a Shiver: Global Warming Protest Frozen Out by Massive Snowfall," *Fox News*, Mar. 2, 2009, http://www.foxnews.com/politics/2009/03/02/shiver-global-warming-protest-frozen-massive-snowfall/

40. Meyers (See no. 38)

41. "WWF and The Coca-Cola Company Team Up to Protect Polar Bears," World Wildlife Foundation, http://www.worldwildlife.org/what/partners/wwfandcoke.html

42. "Acid Test: The Global Challange of Ocean Acidification," NRDC Documentary, Sep. 17, 2009, http://www.youtube.com/watch?v=5cqCvcX7buo

43. Gretchen Hofmann et al., "High-Frequency Dynamics of Ocean pH: A Multi-Ecosystem Comparison," *PLoS ONE*, Dec. 2011, v. 6, Iss. 12, http://www.plosone.org/article/info%3Adoi%2F10.1371%2Fjournal.pone.0028983

44. "Ocean acidification due to increasing atmospheric carbon dioxide," The Royal Society, June 2005, www.royalsoc.ac.uk

45. James Hansen, "Global Warming Twenty Years Later: Tipping Points Near," address to the National Press Club, June 23, 2008, http://www.columbia.edu/~jeh1/2008/TwentyYearsLater_20080623.pdf

46. Glenn Dea'th et al., "Declining Coral Calcification on the Great Barrier Reef," *Science*, v. 2, Jan. 2, 2009, pp. 116-119

47. *CO_2, Global Warming and Coral Reefs: Prospects for the Future* by Craig Idso, (Center for the Study of Carbon Dioxide and Global Change and the Science and Public Policy Institute, 2009)

48. Stephen McIntyre, "'Unprecedented' in the past 153 years," June 3, 2009, http://www.climateaudit.org/?p=6189

49. Byrne et al., "Direct observation of basin-wide acidification of the North Pacific Ocean," Geophysical Research Letters, Jan. 20, 2010, v. 37, L02601

50. Hofmann (See no. 43)

51. Photographs by Bob Halsted

52. Peter Doran and Maggie Zimmerman, "Examining the Scientific Consensus on Climate Change," *EOS Transactions*, v. 90, no. 3, Jan. 20, 2009, pp. 21-22

53. Naomi Oreskes, "The Scientific Consensus on Climate Change," *Science*, v. 306, Dec. 3, 2004, p. 1686

54. Brian Montopoli, "Scott Pelley and Catherine Herrick on Global Warming Coverage," *CBS News*, March 23, 2006, http://www.cbsnews.com/8301-500486_162-1431768-500486.html

55. "The Heidelberg Appeal," American Policy Center, http://www.americanpolicy.org/2002/03/29/the-heidelberg-appeal/

56. "The Leipzig Declaration," American Policy Center, http://www.americanpolicy.org/2002/03/29/the-leipzig-declaration/

57. "The Climate Scientists Register," International Climate Science Coalition, http://www.climatescienceinternational.org/index.php?option=com_content&view=article&id=289&Itemid=17

58. "Global Warming Petition Project," Oregon Institute of Science and Medicine, http://www.petitionproject.org/

59. Brainy Quote, http://www.brainyquote.com/quotes/authors/v/vladimir_lenin.html

60. Nicholas Ballasy, "US Education Secretary Vows to Make American Children 'Good Environmental Citizens,'" *CNS News*, Sep. 23, 2010, http://cnsnews.com/node/75711

61. Penny Starr, "US Education Dep't Pushes Man-Made Global Warming, Saving the Earth at Children's Reading Event," *CNS News*, Aug. 2, 2011, http://cnsnews.com/news/article/us-education-dep-t-pushes-man-made-global-warming-saving-earth-children-s-reading-event

62. "MSU partners bring climate change curricula to high schools," Mississippi State University press release, Feb. 1, 2011, http://www.msstate.edu/web/media/detail.php?id=5134

63. "Climate Kids: NASA's Eyes on the Earth," National Aeronautic and Space Administration, http://climate.nasa.gov/kids/

64. *A Hot Planet Needs Cool Kids: Understanding Climate Change and What You Can Do About It* by Julie Hall (Green Goat Books, 2007)

65. "Act on CO2," YouTube, http://www.youtube.com/results?search_query=act+on+co2&oq=act+on+co2&aq=f&aqi=g3&aql=&gs_sm=e&gs_upl=2137l4803l0l5077l10l7l0l1l1l0l216l779l1.2.2l5l0

66. Anthony Watts, "TV network Tells Kids How Long Their Carbon Footprint Should Allow them to Live," *WattsUpWithThat*, May 31, 2008, http://wattsupwiththat.com/2008/05/31/tv-network-tells-kids-when-their-carbon-footprint-says-they-should-die/

67. Act on CO_2 (See no. 65)

Chapter 9: Bad Science—Temperature, the IPCC, and Revealing E-mails

1.   Henry Lamb, "Kyoto Report," Eco-Logic, Nov/Dec, 1997, http://sovereignty.net/p/clim/kyotorpt.htm

2.   Rachel Waters, "Global Warming Movie Makes the Media Hot for Al Gore All Over Again," *Business and Media Institute*, Aug. 16, 2006, http://www.mrc.org/bmi/reports/2006/Summer_Rerun.html

3.   Michael Hawthorne, "Blunt answers about the risks of global warming," *Chicago Tribune*, Aug. 3, 2008, http://articles.chicagotribune.com/2008-08-03/news/0808020393_1_global-warming-climate-change-rajendra-pachauri

4.   Barack Obama, Message for the Global Climate Summit, November 17, 2008

5.   Robin Bravender, "EPA Chief Goes Toe-To-Toe with Senate GOP Over Climate Science," *New York Times*, Feb. 23, 2010,

http://www.nytimes.com/gwire/2010/02/23/23greenwire-epa-chief-goes-toe-to-toe-with-senate-gop-over-72892.html

6. *The Quotable Einstein* by Alice Calaprice (Princeton University Press, 1996)

7. James Lewis, "Is there an average global temperature," *American Thinker*, Mar. 18, 2007, http://www.americanthinker.com/2007/03/is_there_an_average_global_tem_1.html

8. Joseph D'Aleo and Anthony Watts, "Surface Temperature Records: Policy Driven Deception," Science & Public Policy Institute, Aug. 27, 2010, http://scienceandpublicpolicy.org/originals/policy_driven_deception.html

9. Climate Reference Network (CRN) Site Information Handbook, National Oceanic and Atmospheric Administration, Dec. 2002, http://www1.ncdc.noaa.gov/pub/data/uscrn/documentation/program/X030FullDocumentD0.pdf

10. Anthony Watts, "Is the US Surface Temperature Record Reliable," *SurfaceStations.org*, http://wattsupwiththat.files.wordpress.com/2009/05/surfacestationsreport_spring09.pdf

11. Ibid

12. US State Climate Extremes Committee, National Oceanic and Atmospheric Administration, 2011, http://www.ncdc.noaa.gov/extremes/scec/searchrecs.php

13. "Climate Change: Improvements Need to Clarify National Priorities and Better Align Them with Federal Funding Decisions," Government Accountability Office, May 2011, http://www.gao.gov/new.items/d11317.pdf

14. NOAA (See no. 12)

15. C. Williams, et al., "United States Historical Climatology Network Monthly Temperature and Precipitation Data," NOAA, May 2008, http://www.sustainableoregon.com/webpages/ndp019.html

16. Ibid

17. Ken Stewart, "The Australian Temperature Record—The Big Picture," July 27, 2010, http://wattsupwiththat.com/2010/07/27/the-australian-temperature-record-the-big-picture/

18. Bryan Leyland, New Zealand Climate Science Coalition, 2010, http://nzclimatescience.net/images/PDFs/app3.graph.pdf

19. "New Zealand—Unaffected by Global Warming," New Zealand Climate Science Coalition, July, 2010, http://nzclimatescience.net/index.php?option=com_content&task=view&id=769&Itemid=32

20. Muir Russell, "The Independent Climate Change Emails Review," July, 2010, http://www.cce-review.org/pdf/FINAL%20REPORT.pdf

21. Richard Treadgold, "Are We Feeling Warmer Yet," New Zealand Climate Science Coalition, Nov. 25, 2009, http://www.climateconversation.wordshine.co.nz/docs/awfw/are-we-feeling-warmer-yet.htm

22. "NIWA sued over data accuracy," *Stuff.co.nz*, Aug., 15, 2010, http://www.stuff.co.nz/national/4026330/Niwa-sued-over-data-accuracy

23. John O'Sullivan, "Legal Defeat For Climatologists In Kiwigate Scandal," *Canada Free Press*, Oct. 8, 2010, http://www.canadafreepress.com/index.php/article/28527

24. Andrei Illarionov, "New Study: Hadley Center and CRU Apparently Cherry-picked Russia's Climate Data," Cato Institute, Dec. 17, 2009, http://www.cato-at-liberty.org/new-study-hadley-center-and-cru-apparently-cherry-picked-russias-climate-data/

25. Ibid

26. D' Aleo (See no.8)

27. Intergovernmental Panel on Climate Change, http://www.ipcc.ch/organization/organization.shtml

28. Ibid

29. Stephen Schneider, "Don't Bet All Environmental Changes Will Be Beneficial," *APS Online*, Aug./Sep. 1996, http://www.aps.org/publications/apsnews/199608/environmental.cfm

30. IPCC (See no. 27)

31. Rick Piltz, "House votes 244-179 to kill US funding of IPCC," *ClimateScienceWatch*, Feb. 19, 2011, http://www.climatesciencewatch.org/2011/02/19/house-votes-244-179-to-kill-u-s-funding-of-ipcc/

32. John Zillman, "The IPCC: A View from the Inside," http://www.apec.org.au/docs/zillman.pdf

33. Paul Georgia, "IPCC report criticized by one of its lead authors," The Heartland Institute, June 1, 2001, http://news.heartland.org/newspaper-article/2001/06/01/ipcc-report-criticized-one-its-lead-authors

34. S. Fred Singer, "Good bye, Kyoto," *American Thinker*, March 13, 2011, http://www.americanthinker.com/2011/03/good_bye_kyoto.html

35. Nikhil Lakshman, "The science is absolutely first rate," Rediff.com India News, June 5, 2007, http://www.rediff.com/news/2007/jun/05inter.htm

36. Intergovernmental Panel on Climate Change, Fourth Assessment Report, Working Group II, Chapter 10, (2007), p. 493

37. Gordon Brown presentation at the Major Economies Forum in London, Oct. 2009, http://climatequotes.com/2010/01/27/gordon-brown-claimed-glaciers-could-disappear-entirely-in-25-years/

38. Pallava Bagla, "Himalaya glacier deadline 'wrong,'" *BBC News*, Dec. 5, 2009, http://news.bbc.co.uk/2/hi/8387737.stm

39. David Rose, "Glacier scientist: I knew data hadn't been verified," *Daily Mail*, Jan. 24, 2010, http://www.dailymail.co.uk/news/article-1245636/Glacier-scientists-says-knew-data-verified.html

40. Ibid

41. Intergovernmental Panel on Climate Change, Fourth Assessment Report, Working Group II, Chapter 13, (2007), p. 596

42. Christopher Booker, "Amazongate: At last we reach the source," *Telegraph*, Jul. 10, 2010, http://www.telegraph.co.uk/comment/columnists/christopherbooker/7883372/Amazongate-At-last-we-reach-the-source.html

43. Intergovernmental Panel on Climate Change, Fourth Assessment Report, Synthesis Report, (2007), p. 50

44. "And now for Africagate," EU Referendum, Feb. 7, 2010, http://eureferendum.blogspot.com/2010/02/and-now-for-africagate.html

45. Donna Laframboise, "UN's Climate Bible Gets 21 'F's on Report Card," http://www.noconsensus.org/ipcc-audit/findings-main-page.php

46. Anthony Watts, "The UN 'disappears' 50 million climate refugees, then botches the disappearing attempt," *WattsUpWithThat*, Apr. 15, 2011, http://wattsupwiththat.com/2011/04/15/the-un-disappears-50-million-climate-refugees-then-botches-the-disappearing-attempt/

47. Roy Spencer, "Climategate 2.0: Bias in Scientific Research," Nov. 23, 2011, http://www.drroyspencer.com/

48. E-mail from Phil Jones to Raymond Bradley, Malcolm Hughes, and Michael Mann, Feb. 21, 2005, "The Climategate Emails," Lavoisier Group, March 2010, http://www.lavoisier.com.au/articles/greenhouse-science/climate-change/climategate-emails.pdf

49. E-mail from Michael Mann to Raymond Bradley, Keith Briffa, Tom Crowley, Phil Jones, Michael Mann, Michael Oppenheimer, Jonathan Overpeck, Kevin Trenbreth, and Tom Wigley, June 4, 2003, Lavoisier Group (See no. 48)

50. E-mail from Chris Folland to Keith Briffa, Phil Jones, and Michael Mann, Sep. 22, 1999, Lavoisier Group (See no. 48)

51. E-mail from Keith Briffa to Chris Folland, Phil Jones, and Michael Mann, Sep. 22, 1999, Lavoisier Group (See no. 48)

52. E-mail from Chris Folland to Keith Briffa, Phil Jones, and Michael Mann, Sep. 22, 1999, Lavoisier Group (See no. 48)

53. E-mail from Michael Mann to Keith Briffa, Chris Folland, and Phil Jones, Sep. 22, 1999, Lavoisier Group (See no. 48)

54. E-mail from Phil Jones to Raymond Bradley, Keith Briffa, Malcolm Hughes, Michael Mann, and Tim Osborn, Nov. 16, 1999, Lavoisier Group (See no. 48)

55. E-mail from Tim Osborn to Thomas Kleinen, Dec. 20, 2006, http://www.ecowho.com/foia.php

56. Intergovernmental Panel on Climate Change, Third Assessment Report, Working Group I, Chapter 2, (2001), p. 134

57. Stephen McIntyre, "IPCC and the 'Trick,'" *Climate Audit*, Dec. 10, 2009, http://climateaudit.org/2009/12/10/ipcc-and-the-trick/

58. Al Revere, "An interview with accidental movie star Al Gore," *Grist*, May 9, 2006, http://www.grist.org/article/roberts2

59. "CRU Data Availability," Climate Research Unit, http://www.cru.uea.ac.uk/cru/data/availability/, Dog with bone image by Shane Adams under Creative Commons Attribution Generic License

60. Science and Technology Committee Announcement, UK Parliament, Dec. 7, 2009, http://www.parliament.uk/business/committees/committees-archive/science-technology/s-t-pn04-091207/

61. "The disclosure of climate data from the Climatic Research Unit at the University of East Anglia," House of Commons Science and Technology Committee, Mar. 24, 2010, http://www.publications.parliament.uk/pa/cm200910/cmselect/cmsctech/387/387i.pdf

62. E-mail from Phil Jones to Michael Mann, July 8, 2004, Lavoisier Group (See no. 48)

63. House of Commons (See no. 61)

64. "CRU Scientific Assessment Panel Announced," University of East Anglia, Mar. 22, 2010, http://www.uea.ac.uk/mac/comm/media/press/CRUstatements/SAPannounce

65. "Report of the International Panel set up by the University of East Anglia to examine the research of the Climatic Research Unit," Apr. 12, 2010, http://www.uea.ac.uk/mac/comm/media/press/CRUstatements/SAP

66. Steve McIntyre, "Oxburgh and the Jones Admission," *Climate Audit*, July 1, 2010, http://climateaudit.org/2010/07/01/oxburgh-and-the-jones-admission/

67. E-mail from Phil Jones to Michael Mann, May 29, 2008, Lavoisier Group (See no. 48)

68. Ross McKitrick, "Understanding the Climategate Inquiries," Sep. 2010, http://www.rossmckitrickcomuploads/4/8/0/8/4808045/rmck_climategate.pdf

69. Ibid

70. Ibid

71. Ibid

72. E-mail from Tom Wigley to Phil Jones, Sep. 28, 2008, Lavoisier Group (See no. 48)

73. Muir Russell et al., "The Independent Climate Change Emails Review," July, 2010, www.cce-review.org/pdf/FINAL%20REPORT.pdf

74. Ibid

75. McKitrick (See no. 68)

76. McKitrick (See no. 68)

77. Context statement in second release of Climategate e-mails, Nov. 22, 2011, http://www.ecowho.com/foia.php

78. E-mail from Michael Mann to Phil Jones, Aug. 3, 2004, http://www.ecowho.com/foia.php

79. E-mail from Michael Mann to Phil Jones, May 30, 2008, http://www.ecowho.com/foia.php

80. E-mail from Phil Jones to Thomas Stocker, May 12, 2009, http://www.ecowho.com/foia.php

81. David Rose, "BBC Sought Advice from Global Warming Scientists on Economy, Drama, Music…and Even Game Shows," *Daily Mail*, Nov. 27, 2011, http://www.dailymail.co.uk/news/article-2066706/BBC-sought-advice-global-warming-scientists-economy-drama-music--game-shows.html

82. John Holdren, testimony before the US House Select Committee on Energy Independence and Global Warming, Dec. 2, 2009, http://pjmedia.com/blog/climategate-obamas-science-adviser-confirms-the-scandal-%E2%80%94-unintentionally/

83. E-mail from Mike Hulme to Simon Turok, Feb. 25, 2002, http://www.ecowho.com/foia.php

84. Peter Sissons, "Peter Sissons: I drove out of Television Centre for the final time last month…and I don't have a pang of

regret," *Daily Mail*, July 15, 2009, http://www.dailymail.co.uk/news/article-1199006/PETER-SISSONS-I-drove-Television-Centre-final-time-month--I-dont-pang-regret.html

85. "Securing a Clean Energy Future, the Australian Government's Climate Change Plan," 2001, http://www.cleanenergyfuture.gov.au/clean-energy-future/securing-a-clean-energy-future/

86. "Climate Change Scoping Plan," The California Air Resources Board for the State of California, December, 2008, http://arb.ca.gov/cc/scopingplan/document/scopingplandocument.htm

87. "The UK Low Carbon Transition Plan," Jul. 15, 2009, http://www.decc.gov.uk/publications/basket.aspx?FilePath=White+Papers%2fUK+Low+Carbon+Transition+Plan+WP09%2f1_20090724153238_e_%40%40_lowcarbontransitionplan.pdf&filetype=4#basket

88. "Climate Change Plan for Canada," 2003, http://publications.gc.ca/site/eng/249152/publication.html

89. "National Climate Change Program," China National Development and Reform Commission, June 4, 2007, http://www.china.org.cn/english/environment/213624.htm

90. "Climate Change Indicators in the United States," Environmental Protection Agency, 2010, http://www.epa.gov/climatechange/indicators.html

91. Rod McGuirk, "Study says global warming shrinks birds," *The Christian Science Monitor*, Aug. 21, 2009, http://www.csmonitor.com/Environment/Global-Warming/2009/0821/study-says-global-warming-shrinks-birds

92. "Bigger birds in Central California, courtesy of global climate change," *SF State News*, Oct. 31, 2011, http://www.sfsu.edu/~news/2011/fall/31.html

Chapter 10: Fund My Climate Study

1. "Eisenhower's Farewell Address to the Nation," Jan. 17, 1961, http://mcadams.posc.mu.edu/ike.htm

2. Monika Kopacz, letter to *The New York Times*, Apr. 12, 2009, http://www.nytimes.com/2009/04/12/magazine/12letters-t-THECIVILHERE_LETTERS.html

3. "Advice to the New Administration and Congress: Actions to make our nation resilient to severe weather and climate change," University Corporation for Atmospheric Research and other organizations, http://www.ucar.edu/td/

4. "Climate Change: Improvements Needed to Clarify National Priorities and Better Align Them with Federal Funding Decisions," US Government Accountability Office, May 2011, http://www.gao.gov/products/GAO-11-317

5. Ibid

6. Ibid

7. "UI gets grant to study climate change," *Lewiston Morning Tribune*, Feb. 19, 2011, http://smart-grid.tmcnet.com/news/2011/02/19/5324934.htm

8. Advice (See no. 3)

9. Simon Lewis, "The Met Office's £33 million supercomputer that keeps Britain - and the world - turning…Now try complaining about the forecast," *Daily Mail*, Feb. 17, 2010, http://www.dailymail.co.uk/home/moslive/article-1249957/The-Met-Offices-33-million-supercomputer-keeps-Britain--world--turning--Now-try-complaining-forecast.html

10. "Granthams to fund Institute for Climate change at Imperial College London," Imperial College London News Release, Feb. 26, 2007, http://www3.imperial.ac.uk/newsandeventspggrp/imperialcollege/newssummary/news_26-2-2007-11-56-9

11. Nicola Jones, "Freak weather could have been predicted," *Nature*, Dec. 30, 2010, http://www.nature.com/news/2010/101230/full/news.2010.685.html

12. "Purdue wins $5M global warming crop-research grant," *Associated Press*, Jun. 29, 2011, http://www.ibj.com/article/print?articleId=28010

13. Steven Lawrence, "Climate Change: The US Foundation Response," The Foundation Center, Feb. 2010, http://foundationcenter.org/gainknowledge/research/pdf/researchadvisory_climate.pdf

14. Darren Samuelsohn "Green donors taking time to soul search," *Politico*, May 31, 2011, http://www.politico.com/news/stories/0511/55980.html

15. Joe Romm, "Dirty Money: Big Oil and corporate polluters spent over $500 million to kill climate bill, push offshore drilling," *Climate Progress*, Sep. 27, 2010, http://thinkprogress.org/romm/2010/09/27/206784/dirty-money-oil-companies-special-interest-polluters-spend-millions-to-kill-climate-bil/

16. Steven Lawrence (See no. 13)

17. "Solyndra files for bankruptcy, looks for buyer," *Business Week*, Sep. 6, 2011, http://www.businessweek.com/ap/financialnews/D9PJ89JG0.htm; Burning wind turbine photograph by Polizei Stade

18. *Climatism! Science, Common Sense, and the 21st Century's Hottest Topic* by Steve Goreham (New Lenox Books, 2010), pp. 275-285

19. US Department of Energy: Loan Programs Office, https://lpo.energy.gov/?page_id=45

20. James Kanter, "Banks Urging US to Adopt the Trading of Emissions," *New York Times*, Sep. 26, 2007, http://www.nytimes.com/2007/09/26/business/26bank.html

21. John Roberts, "Science for Sale," *Chemical and Engineering News*, Mar. 24, 2008, http://pubs.acs.org/cen/books/86/8612books.html

22. "Commitments Totaling $6.2 Billion Announced on Day One of Clinton Global Initiative," *Philanthropy News Digest*, Sep. 22, 2011, http://foundationcenter.org/pnd/news/story.jhtml?id=354400023

23. Anita Pugliese and Julie Ray, "Fewer Americans, Europeans View Global Warming as a Threat, Gallup, Apr. 20, 2011, http://www.gallup.com/poll/147203/fewer-americans-europeans-view-global-warming-threat.aspx

24. "Climate change deniers," David Suzuki Foundation, http://www.davidsuzuki.org/issues/climate-change/science/climate-change-basics/climate-change-deniers/

25. William Reilly, "Analysis: UN calls climate debate 'over,'" *UPI*, May 10, 2007, http://www.ecoearth.info/shared/reader/welcome.aspx?linkid=74756&keybold=natural%20AND%20%20resource%20AND%20%20over-use

26. Eoin O'Carroll, "Are Climate-Change Deniers Guilty of Treason?," *Christian Science Monitor*, July 10, 2009, http://www.alternet.org/environment/141204/are_climate-change_deniers_guilty_of_treason/?page=entire; Executioner photo by Andrew Butko under GFDL

27. David Roberts, "An excerpt from a new book by George Monbiot," *Grist*, Sep. 19, 2006, http://www.grist.org/article/the-denial-industry

28. Sharon Begley, "The Truth About Denial," *Newsweek*, Aug. 12, 2007, http://www.thedailybeast.com/newsweek/2007/08/13/the-truth-about-denial.html

29. Gene Ananth, "Will the real ClimateGate please stand up? (Part 2)," Greenpeace, Apr. 1, 2010, http://www.greenpeace.org/international/en/news/Blogs/climate/will-the-real-climategate-please-stand-up-par/blog/9245/;Executioner photo by Andrew Butko under GFDL

30. Data from annual reports or websites of Nature Conservancy, World Wildlife Fund, Environmental Defense Fund, National Resources Defense Council, Sierra Club, Heritage Foundation, American Enterprise Institute, Cato Institute, and Heartland Institute, 2010

31. Anthony Watts, "The great big map of FUD," *WattsUpWithThat*, Oct. 6, 2011, http://wattsupwiththat.com/2011/10/06/the-great-big-map-of-fud/

32. "Gore decries 'global warming' in bitterly cold NYC," *WND*, Jan. 15, 2004, http://www.wnd.com/2004/01/22790/

33. Senator James Inhofe, speech to the US Senate, July 28, 2003, http://inhofe.senate.gov/pressreleases/climate.htm

34. Bill Gray, "On the Hijacking of the American Meteorological Society," *WattsUpWithThat*, Jun. 16, 2011, http://wattsupwiththat.com/2011/06/16/on-the-hijacking-of-the-american-meteorological-society-ams/

35. Steve Zwick, "A Tennessee Fireman's Solution to Climate Change," *Forbes*, Apr. 19, 2012, http://www.forbes.com/sites/stevezwick/2012/04/19/a-tennessee-firemans-solution-to-climate-change/

36. Maibach, E., et al., "A National Survey of Television Meteorologists about Climate Change: Preliminary Findings," George Mason University, Center for Climate Change Communication, 2010, http://www.climateaccess.org/category/organization/george-mason-university-center-climate-change-communication

37. "Open Letter to the Council of the American Physical Society," *Climate Physics*, Sep. 19, 2009, http://climatephysics.org/open-letter-to-the-council-of-the-american-physical-society/

38. Tom Nordlie, "University of Florida-led teams awarded $6.9 million for climate change projects," University of Florida News, June 30, 2011, http://news.ufl.edu/2011/06/30/climate-change-grants/

39. Hal Lewis resignation letter from the American Physical Society, Oct. 6, 2010, http://wattsupwiththat.com/2010/10/16/hal-lewis-my-resignation-from-the-american-physical-society/

40. Lubos Motl, "Nobel Prize Winner Ivar Giaever Resigns Form American Physical Society Over Global Warming Alarmism," Global Warming Policy Foundation, Sep. 14, 2011, http://thegwpf.org/science-news/3870-nobel-prize-winner-ivar-giaever-resigns-from-aps-over-global-warming-alarmism.html

41. Brainy Quote, http://www.brainyquote.com/quotes/authors/u/upton_sinclair.html

Chapter 11: Sunbeams, Zephrys, and Green Leafy Fuel

1.   "Dirty Harry," IMDb, 1971, http://www.imdb.com/title/tt0066999/

2.   "Key Energy World Statistics," International Energy Agency, 2011, http://www.iea.org/textbase/nppdf/free/2011/key_world_energy_stats.pdf

3.   *The Skeptical Environmentalist: Measuring the Real State of the World* by Bjorn Lomborg (Cambridge University Press, 2001), p. 122

4.   "Automobile Industry Introduction," Plunkett Research, Ltd., 2011, http://www.plunkettresearch.com/automobiles-trucks-market-research/industry-overview

5.   Key Energy Statistics (See no. 2)

6.   Key Energy Statistics (See no. 2)

7.   "Chemical Potential Energy," The Physics Hypertextbook, 2011, http://physics.info/energy-chemical/

8.   Ibid

9.   "The Plowboy Interview with Amory Lovins," *Mother Earth News*, Nov.-Dec. 1977, http://www.motherearthnews.com/Renewable-Energy/1977-11-01/Amory-Lovins.aspx?page=14

10. *Earth in Balance: Ecology and the Human Spirit* by Al Gore (Houghton Mifflin, 1992), p. 243

11. Donella Meadows et al., "The Limits to Growth: A Report to the Club of Rome," 1972, http://clubofrome.at/archive/limits.html

12. International Conference on the Future of Energy and the Interconnected Challenges of the 21st Century, The Club of Rome, October, 2011, http://www.clubofrome.org/?p=3086

13. *Rays of Hope: The Transition to a Post-Petroleum World* by Denis Hayes, (Norton, 1977), p. 155
14. Denis Hayes, "Sunpower," *Patagonia*, Winter 2005, http://www.patagonia.com/us/patagonia.go?assetid=2069
15. "Sen. Charles Schumer touts tax incentives for manure energy," *Associated Press*, Aug. 12, 2011, http://www.syracuse.com/news/index.ssf/2011/08/sen_charles_schumer_touts_tax.html; Hamster image by Doenertier82 under Creative Commons License
16. Sam Shrank and Farhad Farahmand, "Biofuels Regain Momentum," Worldwatch Institute, Aug. 29, 2011, http://vitalsigns.worldwatch.org/vs-trend/biofuels-regain-momentum
17. *Climatism! Science, Common Sense, and the 21st Century's Hottest Topic* by Steve Goreham, (New Lenox Books, 2010), p. 332
18. 160,000 derived from IEA Wind 2010 Annual Report, International Energy Agency, July 2011, http://www.ieawind.org/
19. Alexander Neubacher, "Solar Subsidy Sinkhole: Re-Evaluating Germany's Blind Faith in the Sun," *Spiegel On Line*, Jan. 18, 2012, http://www.spiegel.de/international/germany/0,1518,809439,00.html
20. "Integrated Energy Report," California Energy Commission, Nov. 2005, http://www.energy.ca.gov/2005publications/CEC-100-2005-007/CEC-100-2005-007-CMF.PDF
21. Rory Cooper, "Energy Secretary Chu Embraces High Gas Prices, Again," Heritage Foundation, Mar. 21, 2011, http://blog.heritage.org/2011/03/21/energy-secretary-chu-embraces-high-gas-prices-again/
22. Walter Mossberg, "Solar Power Seen Meeting 20% of Needs by 2000; Carter May Seek Outlay Boost," *Wall Street Journal*, Aug. 22, 1978; Crystal ball image by Eva Kröcher under GFDL
23. *The Solar Fraud: Why Solar Energy Won't Run the World* by Howard Hayden (Vales Lake, 2005), p.48
24. Ibid, p. 49
25. "International Energy Outlook 2011," Energy Information Agency, Sep. 2011, http://www.eia.gov/forecasts/ieo/index.cfm
26. Estimated from data in Key Energy Statistics (See no. 2)
27. Key Energy Statistics (See no. 2)
28. Key Energy Statistics (See no. 2)
29. Calculated from Key Energy Statistics reports in 2008 and 2011, International Energy Agency, http://www.iea.org/
30. Key Energy Statistics (See no. 2)
31. Key Energy Statistics (See no. 2)
32. "Monthly Energy Review December 2011," Energy Information Agency, http://www.eia.gov/totalenergy/reports.cfm; *Power Hungry: The Myths of Green Energy and the Real Fuels of the Future* by Robert Bryce (Public Affairs, 2010)
33. Ibid
34. Ralph Nader, 2002 presentation at Western Washington University, http://westernfrontonline.net/news/2075-nader-warns-of-everyday-corporate-control
35. "How Solar Energy Works," Union of Concerned Scientists, http://www.ucsusa.org/clean_energy/technology_and_impacts/energy_technologies/how-solar-energy-works.html
36. *Blue Planet in Green Shackles* by Václav Klaus, (Competitive Enterprise Institute, 2008), p. 86
37. Nora Méray, "Wind and Gas: Back-up or Back-Out, That is the Question," Clingendael International Energy Programme, Dec. 2011, http://www.europeanenergyreview.eu/site/pagina.php?id=3460
38. Burning wind turbine photograph by Polizei Stade
39. "Design report of flue gas and steam integration of power plant and capture plant including interface list," Kingsnorth Carbon Capture and Storage Project, E.ON, 2007
40. David Adam and Mark Tran, "Kingsnorth power station plans shelved by E.ON," *Guardian*, Oct. 7, 2009, http://www.guardian.co.uk/environment/2009/oct/07/eon-cancels-kingsnorth-power-station
41. James Hylko, "Plant Vogtle Leads the Next Nuclear Generation," *Power Magazine*, Nov. 1, 2009, http://www.powermag.com/nuclear/Plant-Vogtle-Leads-the-Next-Nuclear-Generation_2247.html
42. "FPL's West County Energy Center," Florida Power and Light, http://www.fpl.com/environment/plant/west_county.shtml
43. Concentrating solar energy photo by US Bureau of Land Management
44. "Andasol Solar Power Station, Spain," Power-Technology.com, http://www.power-technology.com/projects/andasolsolarpower/; "Andasol-1," National Renewable Energy Laboratory, http://www.nrel.gov/csp/solarpaces/project_detail.cfm?projectID=3
45. Photo of Kingsnorth plant by David Bowen, 2007, under Creative Commons License; Photo of Andasol-1 plant by BSMPS under Creative Commons License; Photo of Vogtle power plant by the Nuclear Regulatory Commission; Photo of offshore wind farm by London Array, http://www.londonarray.com/; Photo of West County Energy Center from Doug Murray, Florida Power and Light
46. London Array, http://www.londonarray.com/
47. "Atlanta Population and Atlanta Demographics, http://www.atlanta.net/visitors/population.html; "Athens," Greek Hotels, http://www.greekhotels.org/greek_islands_area_Athens.html
48. "London, general information," *Have Travel Fun*, http://www.havetravelfun.com/unitedkingdom/london/london-information.htm
49. Wind turbine image by Chrisdesign; Solar panel image by Pirx; Nuclear plant by qubodup
50. "Size of Industrial Wind Turbines," *National Wind Watch*, Jan. 28, 2013, www.wind-watch.org/faq-size-p.php
51. Wind turbine field in Palm Springs photo by Jeff Turner in 2008 under Creative Commons license
52. IEA Wind 2010 Annual Report, International Energy Agency, July 2011, http://www.ieawind.org/

53. "Winter Consultation 2011/2012," UK National Grid, July 2011, http://www.nationalgrid.com/NR/rdonlyres/C3A81245-D988-48A4-80F2-5082F601E06D/48771/WinterConsultation2011PUBLISHV2.pdf

54. Ibid

55. "School to create own wind power," *BBC News*, Feb. 29, 2008, http://news.bbc.co.uk/2/hi/uk_news/england/cornwall/7271533.stm; Burning wind turbine photograph by Polizei Stade

56. "Wind brings down turbine," *Newquay Voice*, Dec. 2, 2009, http://www.newquayvoice.co.uk/news/0/article/2859/

57. UK National Grid (See no. 53)

58. "ERCOT Expects Adequate power Supplies for Summer," ERCOT News Release, May 12, 2010, http://www.ercot.com/news/press_releases/show/329

59. John Sullivan, "Power Supplier Admits Going Green Will Put Lights Out in Britain," *Johnosullivan*, Mar. 5, 2011, http://johnosullivan.livejournal.com/31784.html

60. Kimball Rasmussen, "A Rational Look at Renewable Energy and the Implications of Intermittent Power," Deseret Power, November, 2010, http://www.hwecoop.com/advice/Rational%20Look%20Renewables%201%202.pdf

61. "Wind Capacity Issues," Western Electricity Coordinating Council working paper, Mar. 17, 2010, http://www.wecc.biz/committees/StandingCommittees/JGC/VGS/PWG/VGCapacity031710/Lists/Minutes/Forms/DispForm.aspx?ID=4

62. "20% Wind Energy by 2030: Wind, Backup power, and Emissions," American Wind Energy Association, http://www.awea.org/learnabout/publications/upload/Backup_Power.pdf

63. "How Less Became More: Wind, Power and Unintended Consequences in the Colorado Energy Market," Bentek Energy, Apr. 16, 2010, http://docs.wind-watch.org/BENTEK-How-Less-Became-More.pdf

64. K. de Groot and C. le Pair, "The hidden fuel costs of wind generated electricity," *Windsecret*, http://www.clepair.net/windsecret.html

65. "The Economics of Renewable Energy," House of Lords, Nov. 25, 2008, http://www.publications.parliament.uk/pa/ld200708/ldselect/ldeconaf/195/195i.pdf

66. *Climatism* (See no. 17), p. 275

67. "Levelized Cost of New Generation Resources in the Annual Energy Outlook 2011," Energy Information Agency, Nov. 2010, http://205.254.135.24/oiaf/aeo/electricity_generation.html

68. Ibid

69. Ibid

70. Personal conversation with Chris Namovitz, Energy Information Agency, Jan. 12, 2012

71. "2010 Wind Technologies Market Report," US Department of Energy, June 2011, http://www1.eere.energy.gov/wind/pdfs/51783.pdf

72. "Levelized Cost of New Electricity Generating Technologies," Institute for Energy Research, May 12, 2009

73. Jake Tapper, "Al Gore's 'Inconvenient Truth'?—A $30,000 Utility Bill," *ABC News*, Feb. 26, 2007, http://abcnews.go.com/Politics/GlobalWarming/story?id=2906888&page=1

74. *Ecoscience: Population, Resources, Environment* by Paul Ehrlich et al., (W.H. Freeman, 1970), p. 323

75. Wind Technologies Market Report (See, no. 71)

76. Wind Technologies Market Report (See, no. 71)

77. Andrew Hough "Wind farm revolts blamed for dramatic fall in planning approvals," *Telegraph*, Oct. 28, 2010, http://www.telegraph.co.uk/earth/energy/windpower/8091934/Wind-farm-revolts-blamed-for-dramatic-fall-in-planning-approvals.html

78. "Ethanol Fuel Incentives Applied in the US," California Energy Commission, Jan. 2004, http://www.energy.ca.gov/reports/2004-02-03_600-04-001.PDF

79. "Bush Delivers Speech on Renewable Fuel Sources," *Washington Post*, Apr. 26, 2006, http://www.washingtonpost.com/wp-dyn/content/article/2006/04/25/AR2006042500762.html

80. "Our Energy Future: Creating a Low Carbon Economy," UK Department for Transport and Department for Environment, Food, and Rural Affairs, Feb. 2003, http://www.official-documents.gov.uk/document/cm57/5761/5761.asp

81. "The Way Forward: Strengthening the Transatlantic Partnership," an address by Angela Merkel, Apr. 30, 2007, http://www.acgusa.org/userfiles/%239%20Merkel-Revised.pdf

82. Paul Wescott, "Full Throttle US Ethanol Expansion Faces Challenges Down the Road," US Department of Agriculture, Sep. 2008, http://www.ers.usda.gov/amberwaves/september09/features/ethanolexpansion.htm

83. Directive 2003/30/EC of the European Parliament and of the Council of 8 May 2003 on the promotion of the use of biofuels or other renewable fuels for transport, http://www.bmu.de/english/renewable_energy/doc/print/42460.php

84. "EU Action Against Climate Change: Leading Global Action to 2020 and Beyond," European Commission, 2009, http://ec.europa.eu/clima/publications/docs/post_2012_en.pdf

85. Ronald Steenblik, "Biofuels—At What Cost?," International Institute for Stainable Development, Sep. 2007, http://www.iisd.org/gsi/sites/default/files/oecdbiofuels.pdf

86. Energy Information Agency, US Department of Energy, http://www.eia.gov/

87. Bruno Waterfield, "EU to ban cars from cities by 2050," *Telegraph*, Mar. 28, 2011, http://www.telegraph.co.uk/motoring/news/8411336/EU-to-ban-cars-from-cities-by-2050.html

88. "Technology Roadmap: Biofuels for Transport," International Energy Agency, 2011, http://www.iea.org/papers/2011/biofuels_roadmap.pdf

89. Tiffany Groode and John Heywood, "Biomass to Ethanol: Potential Production and Environmental Impacts," Mass.

Institute of Technology, Feb. 2008, http://web.mit.edu/mitei/lfee/programs/archive/publications/2008-02-rp.pdf

90. "Alligator fat as a new source of biodiesel fuel," *Bio Fuel Daily*, Oct. 28, 2011, http://www.biofueldaily.com/reports/Alligator_fat_as_a_new_source_of_biodiesel_fuel_999.html; Hamster image by Doenertier82 under Creative Commons License

91. Hosain Shapouri et al., "The Energy Balance of Corn Ethanol: An Update," US Department of Agriculture, July 2002, http://www.usda.gov/oce/reports/energy/aer-814.pdf

92. Marcello Oliveira et al., "Ethanol as Fuel: energy, Carbon Dioxide, Balances, and Ecological Footprint," *Bioscience*, July 2005, v. 55, no. 7

93. Groode (See no. 89)

94. Kirk Berge, "Ethanol and Biodiesel: the Good, the Bad, and the Unlikely," Dec. 5, 2008, http://www.peakoil.net/files/Biofuels_%20the%20G_%20the%20B_the%20U.pdf

95. Dennis Avery, "Biofuels, Food, or Wildlife? The Massive Land Costs of US Ethanol," Competitive Enterprise Institute, Sep. 21, 2006, http://cei.org/pdf/5532.pdf

96. Texas windfarm photograph from US Department of Energy

97. "Food Outlook, Global Market Analysis," Food and Agriculture Organization, June 2006, http://www.fao.org/docrep/009/J7927e/J7927e00.htm

98. Biofuels for Transport (See no 88)

99. "Annual Energy Outlook 2011," US Energy Information Adminstration, http://www.eia.gov/forecasts/aeo/er/

100. "Biofuels: Prospects, Risks, and Opportunities," Food and Agriculture Organization of the United Nations, 2008, http://www.fao.org/publications/sofa-2008/en/

101. "US Long Term Corn Projections," US Dept. of Agriculture, Aug. 2011, http://www.usda.gov/wps/portal/usda/usdahome

102. "World Biofuel Production," US Energy Information Adminstration, 2012, http://www.eia.gov/cfapps/ipdbproject/IEDIndex3.cfm; "World Agricultural Outlook Database," Food and Agriculatural Policy Research Institute, Iowa State University, 2012, http://www.fapri.iastate.edu/tools/outlook.aspx

103. Donald Mitchell, "A Note on Rising Food Prices," World Bank, July 2008, http://www-wds.worldbank.org/servlet/WDSContentServer/WDSP/IB/2008/07/28/000020439_20080728103002/Rendered/PDF/WP4682.pdf

104. Lester Brown, "Starving the People to Feed the Cars," *Washington Post*, Sep. 10, 2006, http://www.washingtonpost.com/wp-dyn/content/article/2006/09/08/AR2006090801596_pf.html

105. M. Wu et al., "Consumptive Water Use in the Production of Ethanol and Petroleum Gasoline," Argonne National Laboratory, Jan. 2009, http://www.transportation.anl.gov/pdfs/AF/557.pdf

106. Winnie Gerbens-Leenes et al., "The water footprint of bioenergy," Proceedings of the National Academy of Sciences, v. 106, n. 25, June 23, 2009, http://www.pnas.org/content/106/25/10219.full.pdf+html

107. Wu (See no. 105)

108. Gerbens-Leenes (See no. 106)

109. "Renewable Fuel Standard: Potential Economic and Environmental Effects of US Biofuel Policy," National Research Council of the National Academies, 2011, http://www.nap.edu/catalog.php?record_id=13105

110. "Opinion of the EEA Scientific Committee on Greenhouse Gas Accounting in Relation to Bioenergy," European Environment Agency, Sep. 15, 2011, http://www.euractiv.com/climate-environment/eu-fresh-row-biofuels-green-claims-news-507737

111. Ibid

112. Peter Bella, "Algae for fuel: Obama's next energy plan," *Washington Times*, Feb. 24, 2012, http://communities.washingtontimes.com/neighborhood/middle-class-guy/2012/feb/24/obamas-algae-energy-plan/;Hamster image by Doenertier82 under Creative Commons License

113. "Al Gore's Ethanol Epiphany," *Wall Street Journal*, Nov. 27, 2010, http://online.wsj.com/article/SB10001424052748703572404575634753486416076.html

114. European Commission (See no. 84)

115. "Renewable and Alternative Fuels," US Environmental Protection Agency, 2012, http://www.epa.gov/otaq/fuels/alternative-renewablefuels/index.htm

116. "Presidental memorandum—Federal Fleet Performance," White House press release, May 24, 2011, http://www.whitehouse.gov/the-press-office/2011/05/24/presidential-memorandum-federal-fleet-performance

117. "History of Electric Vehicles," About.com, http://inventors.about.com/od/estartinventions/a/Electric-Vehicles.htm

118. Thomas Edison with electric car image from the Smithsonian

119. Alex Planes, "America's sad love affair with the electric car," *MSN Money*, jan. 6, 2012, http://money.msn.com/investment-advice/latest.aspx?post=b0922a19-be45-45a1-8967-9055b8c94c67

120. John Voelcker, "Obama 2012 Budget Proposes Higher Tax Credit For Plug-In Cars," *The Washington Post*, Feb. 15, 2012, http://www.washingtonpost.com/cars/obama-2012-budget-proposes-higher-tax-credit-for-plug-in-cars/2012/02/15/gIQAPuKgFR_story.html

121. "2017 and Later Model Year Light-Duty Vehicle Greenhouse Gas Emissions and Corporate Average Fuel Economy Standards," Environmental Protection Agency and Department of Transportation, *Federal Register*, v. 76, no. 231, Dec. 1, 2011, http://www.nhtsa.gov/About+NHTSA/Press+Releases/2011/We+Can't+Wait:+Obama+Administration+Proposes+Historic+Fuel+Economy+Standards+to+Reduce+Dependence+on+Oil,+Save+Consumers+Money+at+the+Pump

122. "Electric Vehicles," University of Tennessee Chattanooga Center for Energy, Transportation, and the Environment, 2012, http://www.utc.edu/Research/CETE/electric.php

123. "Useful Conversion Factors," IOR Energy, 2009
124. Bill Canis, "Battery Manufacturing for Hybrid and Electric Vehicles: Policy Issues," Congressional Research Service, Mar. 22, 2011, http://www.fas.org/sgp/crs/misc/R41709.pdf
125. Electric Vehicles (See no. 122)
126. Useful Conversion Factors (See no. 123)
127. "Vapor Recovery Test Procedure," California Environmental Protection Agency, Apr. 12, 1996, http://www.arb.ca.gov/regact/march2000evr/tp-201.5.pdf
128. Canis (See no. 124)
129. William O'Keefe, "Electric Cars: Not Ready for Prime Time," The Marshall Institute, Dec. 2010, http://www.marshall.org/article.php?id=922
130. "Transitions to Alternative Transportation Technologies: Plug-in Hybrid Electric Vehicles," National Academy of Sciences, 2010, http://www.nap.edu/catalog.php?record_id=12826
131. "GM to investigate after Chevy Volt hybrid catches fire for SECOND time in a week—even though it was unplugged." *Daily Mail*, Apr. 19, 2011, http://www.dailymail.co.uk/news/article-1378314/GM-investigate-Chevy-Volt-hybrid-sparks-second-week.html; Burning wind turbine photograph by Polizei Stade
132. *Earth in Balance* (See no. 10)
133. "Electric car owners may face £19,000 battery charge," *Telegraph*, Aug. 1, 2011, http://www.telegraph.co.uk/motoring/news/8674273/Electric-car-owners-may-face-19000-battery-charge.html
134. Sebastian Blanco, "Chevy Volt has best month ever, but Nissan Leaf still wins 2011 plug-in sales contest," Autoblog Green, Jan. 4, 2012, http://green.autoblog.com/2012/01/04/chevy-volt-has-best-month-ever-but-nissan-leaf-still-wins-2011/
135. BrainyQuote.com, Xplore Inc, 2012. http://www.brainyquote.com/quotes/quotes/w/winstonchu135256.html
136. *Climatism* (See no. 17), p. 320
137. "Obama aims for a million green cars by 2015," *ABC News*, Mar. 20, 2009, http://www.abc.net.au/news/2009-03-20/obama-aims-for-a-million-green-cars-by-2015/1625172
138. Leslie Crawford, "Interview transcript: Jose Luis Rodríguez Zapatero," *Financial Times*, June 4, 2008, http://www.ft.com/intl/cms/s/0/7eec944e-324b-11dd-9b87-0000779fd2ac.html#axzz1oitrhT3X
139. Gabriel Calzada Álvarez et al., "Study of the effects on employment of public aid to renewable energy sources," Universidad Rey Juan Carlos, Mar. 2009, http://www.juandemariana.org/pdf/090327-employment-public-aid-renewable.pdf
140. International Energy Annual 2006, US Energy Information Administration, http://205.254.135.7/iea/
141. Asociación de la Industria Fotovoltaica, 2009, http://www.asif.org/
142. "Spain subsidy cuts seen threatening thousands of jobs," *Solar Daily*, Jan. 31, 2012, http://www.solardaily.com/reports/Spain_subsidy_cuts_seen_threatening_thousands_of_jobs_999.html
143. "How to Create a Job: Creating Value, not Just Work," Economic Freedom, Sep. 14, 2009, http://www.economicfreedom.org/2011/09/14/how-to-create-a-job-creating-value-not-just-work/
144. IEA Wind 2010 Annual Report, International Energy Agency, July 2011, http://www.ieawind.org/
145. "Biomass 2020: Opportunities, Challenges, and Solutions," *Eurelectric*, Oct. 2011, http://www2.eurelectric.org/Content/Default.asp?
146. Neubacher (See no. 19)
147. Neubacher (See no. 19)
148. "Economic impacts from the promotion of renewable energies: The German experience," RWI, Oct. 2009, http://www.wind-watch.org/documents/economic-impacts-from-the-promotion-of-renewable-energies-the-german-experience/
149. "Development of renewable energy sources in Germany in 2010," Federal Ministry for the Environment, Nature Conservation and Nuclear Safety, Dec. 2011, http://www.bmu.de/english/renewable_energy/doc/39831.php
150. "The Energy Concept and its accelerated implementation," Federal Ministry for the Environment, Nature Conservation and Nuclear Safety, Oct. 2011, http://www.bmu.de/english/transformation_of_the_energy_system/resolutions_and_measures/doc/48054.php
151. "World in Transition: A Social Contract for Sustainability," The German Advisory Council on Global Change, Mar. 2011, http://www.wbgu.de/fileadmin/templates/dateien/veroeffentlichungen/hauptgutachten/jg2011/wbgu_jg2011_kurz_en.pdf
152. Ibid
153. "Fukushima Investigative Report," Dec, 26, 2011, http://icanps.go.jp/eng/120224SummaryEng.pdf
154. "Germany to Phase Out Nuclear Power by 2022," *Spiegel On Line*, May 30, 2011, http://www.spiegel.de/international/germany/0,1518,765594,00.html
155. Günther Kiel, "Germany's Energy Supply Transformation Has Already Failed," European Institute for Climate and Energy, Dec. 2011, http://www.eike-klima-energie.eu/fileadmin/user_upload/Bilder_Dateien/Keil_Energiewende_gescheitert/2012-EIKE_Germanys_green_energy_turnaround_V2.pdf
156. Development of renewable energy sources (See no. 149)
157. Development of renewable energy sources (See no. 149)
158. Neubacher (See no. 19)
159. RWI (See no. 148)
160. "European Residential Energy Price Report 2011," VaasaETT, Mar. 8,2011, http://www.vaasaett.com/2011/04/european-

residential-energy-price-report-2011-released/

161. "Energy transition affects one in five industry operating activities moved abroad or plan to do so!," European Institute for Climate and Energy, Jan. 20, 2012, http://www.eike-klima-energie.eu/news-cache/energiewende-wirkt-jeder-fuenfte-industriebetrieb-verlegt-aktivitaeten-ins-ausland-oder-plant-dies/

162. Marc Roca and Netalie Pearson, "Solar Stocks Plunge as Germany Vows to Quicken Subsidy Cuts," *Bloomberg Businessweek*, Jan. 20, 2012, http://www.bloomberg.com/news/2012-01-20/solar-stocks-plunge-worldwide-as-germany-accelerates-rate-cuts.html

163. Neubacher (See no. 19)

164. Angelo Racoma, "Solar Cell Prices Stabilize at US$ 0.50 per Watt," *TechwireAsia*, Feb. 20, 2012, http://www.techwireasia.com/2021/solar-cell-prices-stabilize-at-us-0-50-per-watt/

165. Jeremy van Loon, "Merkel's Nuclear Plan Earns Derision as Clean Power Costs Climb," *Bloomberg*, Jan. 4, 2011, http://www.bloomberg.com/news/2011-01-04/merkel-s-nuclear-embrace-earns-derision-as-german-clean-power-costs-climb.html

166. "Campaigners welcome PM's pledge to lead 'greenest Government ever,'" *ClickGreen*, May 15, 2010, http://www.clickgreen.org.uk/news/national-news/121317-eco-sector-welcomes-pms-pledge-to-be-%E2%80%9Cgreenest-government-ever%E2%80%9D.html

167. Andrew Orlowski, "Snow blankets London for Global Warming debate," *Register*, Oct. 29, 2008, http://www.theregister.co.uk/2008/10/29/commons_climate_change_bill/

168. Ibid

169. "UK Energy in Brief 2011," Department of Energy and Climate Change, July 2011, http://www.decc.gov.uk/assets/decc/11/stats/publications/energy-in-brief/2286-uk-energy-in-brief-2011.pdf

170. "Renewables Output in 2010," Renewable Energy Foundation, Apr. 12, 2011, http://www.ref.org.uk/publications/229-renewables-output-in-2010

171. Aaron Patrick, "Wind energy endures a gale of hostility," *Telegraph*, Mar. 26, 2005, http://www.telegraph.co.uk/finance/2912692/Wind-energy-endures-a-gale-of-hostility.html

172. IEA Wind (See no. 144)

173. Renewables Output (See no. 170)

174. Tom Peterkin, "Meeting 2020 renewable energy targets would cost households £4,000 a year," *Telegraph*, June 25, 2008, http://www.telegraph.co.uk/earth/earthnews/3345397/Meeting-2020-renewable-energy-targets-would-cost-households-4000-a-year.html

175. Robert Mendick and Edward Malnick, "The aristocrats cashing in on Britain's wind farm subsidies," *Telegraph*, Aug. 21, 2011, http://www.telegraph.co.uk/earth/energy/windpower/8713128/The-aristocrats-cashing-in-on-Britains-wind-farm-subsidies.html

176. UK Energy in Brief (See no. 169)

177. Rowena Mason, "Government accused of 'destroying 25,000 green jobs,'" *Telegraph*, Oct. 31, 2011, http://www.telegraph.co.uk/news/8859934/Government-accused-of-destroying-25000-green-jobs.html

178. J.P. Lesley, State Geologist of Pennsylvania, 1886, quoted in *Standard Oil Company (Indiana): Oil Pioneer of the Middle West* by Paul Giddens (Appleton-Century-Crofts, 1955), p.2; Crystal ball image by Eva Kröcher under GFDL

179. L. Schneider and B. Brooks, American Association of Petroleum Geologists Bulletin, 1936, quoted in Edward Porter, "Are We Running Out of Oil?," American Petroleum Institute, Discussion paper #81, Dec. 1995

180. *Energy and the Future* by Allen Hammond et al. (American Association for the Advancement of Science, 1973), p.vi

181. *World Energy Strategies: Facts, Issues, and Options* by Amory Lovins (Friends of the Earth International, 1975), p. 26

182. "Conferences: Opening the Debate," *Time Magazine*, Apr. 25, 1977, http://www.time.com/time/magazine/article/0,9171,918861,00.html

183. "The World Oil market in the Years Ahead," US Central Intelligence Agency, Aug. 1979, p. iii

184. Dave Cohen, "The Perfect Storm," Energy Bulletin, ASPO, Oct. 31, 2007, http://www.energybulletin.net/node/36510

185. *Out of Gas: The End of the Age of Oil* by David Goodstein (W.W. Norton, 2004), p. 123

186. "International Energy Statistics," Energy Information Agency, http://www.eia.gov/cfapps/ipdbproject/IEDIndex3.cfm?tid=5&pid=53&aid=1

187. Ibid

188. "A Time Comes: What it means to take action," Greenpeace, June 3, 2009, http://www.greenpeace.org/usa/en/news-and-blogs/news/what-it-means-to-take-action/

189. Sierra Club, http://www.sierraclub.org/

190. Mark Drajem, "NASA's Hansen Arrested Outside White House at Pipeline Protest," *Bloomberg*, Aug. 29, 2011, http://www.bloomberg.com/news/2011-08-29/nasa-s-hansen-arrested-outside-white-house-at-pipeline-protest.html

191. International Energy Statistics (See no. 186)

192. Nathan Myhrvold, "Energy Revolution Keeps Carbon on top: Myhrvold," *Bloomberg*, Oct. 26, 2011, http://www.bloomberg.com/news/2011-10-26/the-energy-revolution-that-keeps-carbon-on-top-nathan-myhrvold.html

193. "The Facts About Fracking," *Wall Street Journal*, June 25, 2011, http://online.wsj.com/article/SB10001424052702303936704576398462932810874.html

194. "A Tale of Two Shale States," *Wall Street Journal*, July 26, 2011, http://online.wsj.com/article/SB10001424052702303678

04576442053700739990.html

195. "World Shale Gas Resources: An Initial Assessment of 14 Regions Outside the United States," Energy Information Administration, Apr. 2011, http://www.eia.gov/analysis/studies/worldshalegas/

196. Greenpeace (2000) quoted in *The Skeptical Environmentalist* by Bjorn Lomborg (Cambridge, 2001), p. 258

197. Andrew Gilligan, "Field of dreams, or an environment nightmare?," *Telegraph*, Nov. 26, 2011, http://www.telegraph.co.uk/earth/energy/gas/8918399/Field-of-dreams-or-an-environment-nightmare.html

198. John Broderick et al., "Shale gas: an updated assessment of environmental and climate change impacts," Tyndall Centre for Climate Change Research, Nov. 2011, http://www.tyndall.ac.uk/

199. "Gasland," IMDb, 2010,http://www.imdb.com/title/tt1558250/

200. Phelim McAleer, "The Gasland movie: a fracking shame—director pulls video to hide inconvenient truths," WattsUpWithThat, June 4, 2011, http://wattsupwiththat.com/2011/06/04/the-gasland-movie-a-fracking-shame-director-pulls-video-to-hide-inconvenient-truths/

201. Thomas Jefferson. BrainyQuote.com, Xplore Inc, 2012. http://www.brainyquote.com/quotes/quotes/t/thomasjeff101820.html, accessed March 10, 2012

202. Dieter Helm, "The peak oil brigade is leading us into bad policymaking on energy," *Guardian*, Oct. 18, 2011, http://www.guardian.co.uk/commentisfree/2011/oct/18/energy-price-volatility-policy-fossil-fuels

203. Energy Information Administration, http://www.eia.gov/

204. Robert Bradley, "Shale Oil vs. Peak Oil," Institute for Energy Research, July 7, 2011, http://www.instituteforenergyresearch.org/2011/07/07/shale-oil-vs-peak-oil/

Chapter 12: You Can't Make This Stuff Up!

1. "International Cremation Statistics 2008," The Cremation Society of Great Britain, http://www.srgw.demon.co.uk/CremSoc5/Stats/Interntl/2008/StatsIF.html

2. Marina Kamenev, "Aquamation: A Greener Alternative to Cremation?," *Time Magazine*, Sep. 28, 2010, http://www.time.com/time/health/article/0,8599,2022206,00.html

3. Holly Richmond, "Crematorium to heat UK swimming pool," *Grist*, Feb. 4, 2011, http://grist.org/living/2011-02-03-crematorium-to-heat-u-k-swimming-pool/

4. "An Introduction to Aquamation," Aquamation Industries, 2010, http://www.aquamationindustries.com/Aquamation%20-%20An%20Introduction.pdf

5. Brid-Aine Parnell, "Global warming COULD SHRINK THE HUMAN RACE," *The Register*, Feb. 24, 2012, http://www.theregister.co.uk/2012/02/24/high_temps_shrink_mammals/

6. "Spanish Solar-Panel Trade Group Calls for Fraud Investigation," *Bloomberg*, Apr. 12, 2010, http://saveourseashore.org/?p=1395

7. "Navy Explores Biofuels," US Navy, http://www.navy.mil/swf/mmu/mmplyr.asp?id=13846

8. James Bartis and Lawrence van Bibber, "Alternative Fuels for Military Applications, Rand Corporation, 2011, http://www.rand.org/pubs/monographs/MG969.html

9. "Operational Energy Strategy: Implementation Plan," Department of Defense, Mar. 2012, http://energy.defense.gov/Operational_Energy_Strategy_Implementation_Plan.pdf

10. Andrew Barr, "Biofuels and the US Navy's 'Great Green Fleet,'" The Heartland Institute, http://blog.heartland.org/2011/12/biofuels-and-the-u-s-navys-great-green-fleet-2/

11. Bartis and Bibber (See no. 8)

12. Andrew Herndon, "Navy to Buy $12 Million of Advanced Biofuels in Record Purchase," *Bloomberg*, Dec. 5, 2011, http://www.bloomberg.com/news/2011-12-05/navy-to-buy-12-million-of-advanced-biofuels-in-record-purchase.html

13. Jim Lane, "US Government to Invest $510 Million in Advanced, Drop-in Biofuels," *Biofuels Digest*, Aug. 16, 2011, http://www.renewableenergyworld.com/rea/news/article/2011/08/us-government-to-invest-510-million-in-advanced-drop-in-biofuels

14. Bartis and Bibber (See no. 8)

15. Karin Zeitvogel, "Global warming could spur toxic algae, bacteria in seas," *Sydney Morning Herald*, Feb. 20, 2011, http://news.smh.com.au/breaking-news-world/global-warming-could-spur-toxic-algae-bacteria-in-seas-20110220-1b0oz.html

16. "Are Scarborough's stop signs contributing to global warming," *Inside Toronto*, Jan. 19, 2012, http://www.insidetoronto.com/news/local/article/1281403

17. Pilita Clark, "Australia poised to allow camel cull," *Financial Times*, June 7, 2011, http://www.ft.com/intl/cms/s/0/6e633ac8-9126-11e0-9668-00144feab49a.html#axzz1pgoJ8GDU

18. "Feral animal management positive list activity," Department of Climate Change and Energy Efficiency, Feb. 8, 2012, http://www.climatechange.gov.au/en/government/initiatives/carbon-farming-initiative/activities-eligible-excluded/additional-activities-positive-list/feral-animal-management.aspx

19. Camel photo by Jjron, Jul. 7, 2007 under GNU Free Documentation License

20. "Developing Methods to Reduce Bird Mortality in the Altamont Pass Wind Resource Area," Pier Report for the California Energy Commissions, Aug. 2004, http://www.energy.ca.gov/reports/500-04-052/500-04-052_00_EXEC_SUM.PDF

21. Albert Manville, "Towers, Turbines, Power Lines, and Buildings—Steps Being Taken by the US Fish and Wildlife Service to

Avoid or Minimize Take of Migratory Birds at These Structures," Proceedings of the Fourth International Flight Conference, 2009, http://utahcbcp.org/htm/tall-structure-info/publication=12537

22. Eagle killed by wind turbine image by Stellan Hedgren
23. "Dorset school's 'bird killer' wind turbine turned off," *BBC News*, July 6, 2010, http://www.bbc.co.uk/news/10518796
24. "Primary school forced to turn off wind turbine after bird deaths," *Telegraph*, July 4, 2010, http://www.telegraph.co.uk/earth/energy/7870929/Primary-school-forced-to-turn-off-wind-turbine-after-bird-deaths.html
25. "Pillar 2: Sustainable Resource Use," Green Schools Initiative, http://greenschools.net/article.php?id=130
26. Kate Loveys, "Foiled by the winter: The £25,000 eco-classroom that can't be used because solar panels don't provide enough heat," *Daily Mail*, Feb. 24, 2011, http://www.dailymail.co.uk/news/article-1360297/25-000-eco-classroom-used-solar-panels-dont-provide-heat.html
27. "Muswell Hill Low Carbon Zone," Haringey Council, http://www.haringey.gov.uk/lcz
28. Living ark image provided by Sue Dunster of Zed Factory Ltd.
29. Foiled by the winter (See no. 26)
30. Michael Zennie, "Students step over 'rivers of urine' after green bathrooms plan for waterless urinals turns a school yellow…and it will cost $500,000 to fix," *Daily Mail*, Jan. 30, 2012, http://www.dailymail.co.uk/news/article-2093979/Students-step-rivers-urine-green-bathrooms-plan-turns-high-school-yellow--cost-500-000-fix.html
31. Southern California wind turbine farm photograph by James McCauley under Creative Commons License
32. "LACCD Builds Green," Los Angeles Community College, http://www.laccdbuildsgreen.org/sustainable-aboutProgram.php
33. Michael Finnegan and Gale Holland, "Grand dream loses sheen in glare of daylight," *Los Angeles Times*, Mar. 6, 2011, http://www.latimes.com/news/local/la-me-build-20110306,0,4909175.story
34. "Energy Oversight Committee Report," Los Angeles Community College, Nov. 18, 2009, http://www.laccdbuildsgreen.org/sustainable-renewablePlan.php
35. "East Los Angeles College - 1 Megawatt Photovoltaic Power Facility," Education Design Showcase, http://www.educationdesignshowcase.com/view.esiml?pid=209
36. Ibid
37. Grand dream (See no. 33)
38. Tiffany Gabbay, "Education British Principal Cuts Off Classrooms' Heat Amid Near-Freezing Temperatures to Lower Carbon Footprint," *Blaze*, Dec. 7, 2011, http://www.theblaze.com/stories/british-principal-cuts-off-clasrooms-heat-amid-near-freezing-temperatures-to-lower-carbon-footprint/
39. "Earth Hour 2012," World Wildlife Fund, http://www.worldwildlife.org/sites/earthhour/index.html
40. "Earth Science Picture of the Day," EPOD, NASA's Earth Science Division, http://epod.usra.edu/blog/2002/02/north-and-south-korea-at-night-1.html
41. "Climate change could be causing cougar attacks: expert," *Canada.com*, Aug. 29, 2007, http://www.canada.com/nationalpost/news/story.html?id=c5e6120a-be10-4497-8f32-cd8585e5ca33&k=51234
42. Agence France-Presse, "Climate change cited as shark attacks 'double,'" *Raw Story*, Feb. 13, 2012, http://www.rawstory.com/rs/2012/02/13/climate-change-cited-as-shark-attacks-double/
43. John Vidal, "There are many tiger widows here," *Guardian*, Sep. 25, 2008, http://www.guardian.co.uk/environment/2008/sep/25/conservation.climatechange
44. Bonnie Malkin, "Duck-billed platypus at risk from climate change," *Telegraph*, June 24, 2011, http://www.telegraph.co.uk/earth/wildlife/8596068/Duck-billed-platypus-at-risk-from-climate-change.html
45. Menchie Tiongan, "Bats Kill More Children in Brazil," *Seer Press News*, Sep. 25, 2010, http://seerpress.com/bats-kill-more-children-in-brazil/7675/
46. Derek Abma, "Global warming, spread of infected ticks linked," *Vancouver Sun*, Mar. 6, 2012, http://www.vancouversun.com/health/Global+warming+spread+infected+ticks+linked/6256364/story.html
47. Eoin O'Carroll, "New hybrid sharks discovered: Signs of global warming?," *Christian Science Monitor*, Jan. 3, 2012, http://www.csmonitor.com/Science/2012/0103/New-hybrid-sharks-discovered-Signs-of-global-warming
48. Zachary Shahan, "Lizard that Outlived Dinosaurs May Go Extinct from Climate Change," *Planetsave*, June 19, 2010, http://planetsave.com/2010/06/19/lizard-that-outlived-dinosaurs-may-go-extinct-from-climate-change/
49. Emma Young, "Global warming poses deaf threat to tropical fish," *New Scientist*, Mar. 6, 2008, http://www.newscientist.com/article/dn13417-global-warming-poses-deaf-threat-to-tropical-fish.html
50. "Flowers losing scent due to climate change," *Asia One*, Mar. 22, 2010, http://news.asiaone.com/News/AsiaOne+News/Malaysia/Story/A1Story20100322-206015.html

Chapter 13: Climatism—Headed for a Crash

1. J.T. Houghton, "Climate Change: The IPCC Scientific Assessment," IPCC Working Group I, 1990, p. XXII
2. David Evans, "The Skeptic's Case," Feb. 27, 2012, http://jonova.s3.amazonaws.com/guest/evans-david/skeptics-case.pdf
3. James Hansen, et. al, "Earth's Energy Imbalance: Confirmation and Implications," *Science*, v. 308, June 3, 2005
4. "Argo: part of the integrated global observation strategy," University of California at San Diego, http://www.argo.ucsd.edu/
5. Houghton (See no. 1)
6. Evans (See no. 2)

7.   Hansen (See no. 3)

8.   Evans (See no. 2)

9.   Kevin Trenberth, "An imperative for climate change planning: tracking Earth's global energy," *ScienceDirect*, v. 1, pp. 19-27, 2009, http://www.cgd.ucar.edu/cas/Trenberth/trenberth.papers/EnergyDiagnostics09final2.pdf

10. David Archibald, "Solar Cycle 24: Expectations and Implications," *Energy & Environment*, v. 20, n. 1&2, 2009, http://www.davidarchibald.info/papers/Archibald2009E&E.pdf

11. "Climate Change 2007: Synthesis Report," Intergovernmental Panel on Climate Change, p. 39, http://www.ipcc.ch/publications_and_data/ar4/syr/en/contents.html

12. Barack Obama speech in St. Paul, Minnesota, June 3, 2008, http://www.youtube.com/watch?v=gl3FLN1t8j0

13. "The American Clean Energy and Security Act (Waxman-Markey Bill)," Center for Climate and Energy Solutions, http://www.c2es.org/federal/congress/111/acesa

14. "Questions and Answers on the Communication Towards a Comprehensive Climate Change Agreement in Copenhagen," Europa.eu, Jan. 28, 2009, http://europa.eu/rapid/pressReleasesAction.do?reference=MEMO/09/34&format=HTML&aged=1&language=EN&guiLanguage=en

15. Fred Pierce, "Earth Summit is doomed to fail, say leading ecologists," *New Scientist*, Feb. 10, 2012, http://www.newscientist.com/article/dn21465-earth-summit-is-doomed-to-fail-say-leading-ecologists.html

16. "China Threatens massive Venting of Super Greenhouse Gases in Attempt to Extort Billions as UNFCCC Meeting Approaches," *PRNewswire*, Nov. 8, 2011, http://www.prnewswire.com/news-releases/china-threatens-massive-venting-of-super-greenhouse-gases-in-attempt-to-extort-billions-as-unfccc-meeting-approaches-133433253.html

17. The Heartland Institute, http://climateconferences.heartland.org/

18. *Climate Change Reconsidered: The Report of the Nongovernmental International Panel on Climate Change* (The Heartland Institute, 2009)

19. "Copenhagen Accord," United Nations Framework Convention on Climate Change, Dec. 18, 2009, http://unfccc.int/files/meetings/cop_15/application/pdf/cop15_cph_auv.pdf

20. Paul Krugman, "Cassandras of Climate," *New York Times*, Sep. 28, 2009, http://www.nytimes.com/2009/09/28/opinion/28krugman.html

21. Dirk Kurbjuweit et. al, "Merkel Abandons Aim of Binding Climate Agreement," *Spiegel Online*, Apr. 26, 2010, http://www.spiegel.de/international/germany/0,1518,691194,00.html

22. "Milestones on the road to 2012: The Cancun Agreements," United Nations Framework Convention on Climate Change, http://unfccc.int/essential_background/cancun_agreements/items/6132.php

23. Arthur Max, "Climate conference approves landmark deal," *Associated Press*, Dec. 11, 2011, http://cnsnews.com/news/article/climate-conference-approves-landmark-deal

24. "Kyoto deal loses four big nations," *Sydney Morning Herald*, May 29, 2011, http://www.smh.com.au/environment/climate-change/kyoto-deal-loses-four-big-nations-20110528-1f9dk.html

25. "Long-Term Trend in Global CO_2 Emissions," PBL Netherlands Environmental Assessment Agency, Sep. 11, 2011, http://www.pbl.nl/en/publications/2011/long-term-trend-in-global-co2-emissions-2011-report

26. Wind turbines in Hawaii image by ©PF Bentley/PFPIX.com

27. Solar farm image by Knut Löschke

28. Emissions (See no. 25)

29. "UK's Carbon Footprint 1990-2009," Department for the Environment, Food, and Rural Affairs, Mar. 8, 2012, http://www.defra.gov.uk/statistics/files/Release_carbon_footprint_08Mar12.pdf

30. "Solar Trust of America files bankruptcy," *Reuters*, Apr. 2, 2012, http://www.reuters.com/article/2012/04/02/us-solartrust-bankruptcy-idUSBRE8310ZV20120402; Burning wind turbine photograph by Polizei Stade"

31. Reducing emissions from transport," European Commission, Jan. 6, 2011, http://ec.europa.eu/clima/policies/transport/index_en.htm

32. "Euro MPs back patio heaters ban," *BBC News*, Jan. 31, 2008, http://news.bbc.co.uk/2/hi/europe/7219565.stm

33. "Report of the World Commission on Environment and Development," United Nations General Assembly, A/RES/42/187, Dec. 11, 1987, http://www.un.org/documents/ga/res/42/ares42-187.htm

34. "The Millinium Development Goals Report," United Nations, 2010, http://www.un.org/millenniumgoals/pdf/MDG%20Report%202010%20En%20r15%20-low%20res%2020100615%20-.pdf

35. "Who's Winning the Clean Energy Race," The Pew Charitable Trusts, 2011, http://www.pewtrusts.org/uploadedFiles/ww-wpewtrustsorg/Reports/Global_warming/G-20%20Report.pdf

36. Organisation for Economic Co-operation and Development (OECD), http://www.oecd.org/home/0,2987,en_2649_201185_1_1_1_1_1,00.html

INDEX

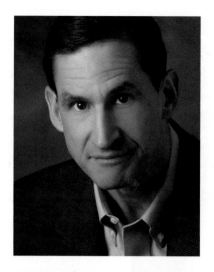

ABOUT THE AUTHOR

Steve Goreham is a speaker, author, and researcher on environmental issues and former engineer and business executive. He is a frequently invited guest on radio and television as well as a freelance writer. He's the Executive Director of the Climate Science Coalition of America (CSCA), a non-political association of scientists, engineers, and citizens dedicated to informing Americans about the realities of climate science and energy economics. CSCA is the US affiliate of the International Climate Science Coalition.

The Mad, Mad, Mad World of Climatism is Steve's second book on climate change, now with over 100,000 copies in print. His first book is *Climatism! Science, Common Sense, and the 21st Century's Hottest Topic* (New Lenox Books, 2010), a complete, in-depth discussion of the science, politics, and energy policy implications of the man-made global warming debate. Steve continues to be astonished every day by unfounded claims of looming global warming catastrophe. He wrote this book to bring the latest facts to the reader, but to also poke fun at a mankind far down the primrose path of global warming fantasy.

Steve holds an MS in Electrical Engineering from the University of Illinois and an MBA from the University of Chicago. He has more than 30 years of experience at Fortune 100 and private companies in engineering and executive roles. He is a husband and father of three and resides in Illinois.